T0137270

Hybrid Metaheuristics for Image Analysis

Siddhartha Bhattacharyya

Editor

Hybrid Metaheuristics for Image Analysis

 Springer

Editor
Siddhartha Bhattacharyya
Department of Computer Applications
RCC Institute of Information Technology
Kolkata
West Bengal, India

ISBN 978-3-030-08497-4   ISBN 978-3-319-77625-5   (eBook)
https://doi.org/10.1007/978-3-319-77625-5

Printed on acid-free paper

This Springer imprint is published by the registered company Springer Nature Switzerland AG.
The registered company address is: Gewerbestrasse 11, 6330 Cham, Switzerland

Siddhartha Bhattacharyya would like to dedicate this book to his late father Ajit Kumar Bhattacharyya, his late mother Hashi Bhattacharyya, his beloved wife Rashni Mukherjee, and his elder cousin sister Piyali Mukherjee

Preface

A metaheuristic is a higher-level procedure designed to select a heuristic (partial search algorithm) that may lead to a sufficiently good solution to an optimization problem, especially with incomplete or imperfect information. The basic principle of metaheuristics is to sample a set of solutions which is large enough to be completely sampled. As metaheuristics make few assumptions about the optimization problem to be solved, they may be put to use in a variety of problems. Metaheuristics do not, however, guarantee that a globally optimal solution can be found on some class of problems since most of them implement some form of stochastic optimization. Hence the solution found is often dependent on the set of random variables generated. By searching over a large set of feasible solutions, metaheuristics can often find good solutions with less computational effort than optimization algorithms, iterative methods, or simple heuristics. As such, they are useful approaches for optimization problems.

Even though the metaheuristics are robust enough to yield optimum solutions, they often suffer from time complexity and degenerate solutions. In an effort to alleviate these problems, scientists and researchers have come up with the hybridization of the different metaheuristic approaches by conjoining them with other soft computing tools and techniques to yield fail-safe solutions. In a recent advancement, quantum mechanical principles are being employed to cut down the time complexity of the metaheuristic approaches to a great extent. Thus, the hybrid metaheuristic approaches have come a long way in dealing with real life optimization problems quite successfully.

Proper and faithful analysis of digital images has been at the helm of affairs in the computer vision research community given the varied amount of uncertainty inherent in digital images. Images exhibit varied uncertainty and ambiguity of information and hence understanding an image scene is far from being a general procedure. The situation becomes even graver when the images become corrupt with noise artifacts. The applications of proper analysis of images encompass a wide range of applications which include image processing, image mining, image inpainting, video surveillance, and intelligent transportation systems to name a few. One of the notable areas of research in image analysis is the estimation of age progression in human beings through analysis of wrinkles in face images, which can be further utilized for tracing unknown or missing persons. Hurdle detection is one of the common tasks in robotic vision that has been done through image processing, by identifying different types of objects in the image and then calculating the distance between the robot and the hurdles. Image analysis has a lot to contribute in this direction.

This volume reports on the latest results or progress in the development of hybrid metaheuristic techniques for faithful image analysis and understanding. The book comprises nine chapters.

The chapter "Current and Future Trends in Segmenting Satellite Images Using Hybrid and Dynamic Genetic Algorithms" presents the foundation of a novel variant of the genetic algorithm named the Hybrid Dynamic Genetic Algorithm. The basis of the hybridization of the proposed genetic algorithm resorts to variable length chromosomes. An application of this algorithm is demonstrated on image segmentation.

In the chapter "A Hybrid Metaheuristic Algorithm Based on Quantum Genetic Computing for Image Segmentation", the authors present a new algorithm for edge detection based on the hybridization of quantum computing and metaheuristics. The main idea is the use of cellular automata as a complex system for image modeling, and the quantum algorithm as a search strategy. The proposed Quantum Genetic Algorithm is found to be effective in edge detection.

The chapter "Genetic Algorithm Implementation to Optimize the Hybridization of Feature Extraction and Metaheuristic Classifiers" presents two face recognition frameworks involving the hybridization of both the feature extraction and classification stages. Feature extraction is performed through the two proposed hybrid techniques, one based on the orthogonal combination of local binary patterns and a histogram of oriented gradients, and the other based on Gabor filters and Zernike moments. A hybrid metaheuristic classifier is also investigated for classification based on the integration of genetic algorithms (GA) and support vector machines (SVM), where GA is used for optimization of the SVM parameters.

The chapter "Optimization of a HMM-Based Hand Gesture Recognition System Using a Hybrid Cuckoo Search Algorithm" focuses on the optimization of a HMM-based hand gesture recognition system using a hybrid cuckoo search algorithm. The authors present a comparative analysis of other classification techniques used in hand gesture recognition with their proposed hybridized bio-inspired metaheuristic approach, namely the Cuckoo Search Algorithm for reducing the complex trajectory in the hidden Markov model (HMM).

In the chapter "Satellite Image Contrast Enhancement Using Fuzzy Termite Colony Optimization", the authors propose the Termite Colony Optimization (TCO) algorithm based on the behavior of termites. Thereafter they use the proposed TCO algorithm and fuzzy entropy for satellite image contrast enhancement. The proposed technique has been found to offer better contrast enhancement of images by utilizing a type-2 fuzzy system and TCO.

The goal of the segmentation techniques called deformable models is to adapt a curve in order to optimize the overlapping with another image of interest with the actual contour. Some of the problems existing in optimization are the choosing of an optimization method, the selection of parameters, and the initialization of curves. The chapter "Image Segmentation Using Metaheuristic-Based Deformable Models" discusses these problems with reference to metaheuristics which are designed to solve complex optimization and machine learning problems.

The chapter "Hybridization of the Univariate Marginal Distribution Algorithm with Simulated Annealing for Parametric Parabola Detection" presents a new hybrid optimization method based on the univariate marginal distribution algorithm for a continuous domain, and the heuristic of simulated annealing for the parabola detection problem. The proposed hybrid method is applied on the DRIVE database of retinal fundus images to approximate the retinal vessels as a parabolic shape. The hybrid method is applied separately using two different objective functions. Firstly, the objective function only considers the superposition of pixels between the target pixels in the input image and the virtual parabola, and secondly, the objective function implements a weighted restriction on the pixels close to the parabola vertex. Both objective functions in the hybrid method obtain suitable results to approximate a parabolic form on the retinal vessels present in the retinal images.

Thresholding is the simplest image segmentation method, where a global or local threshold value is selected for segmenting pixels into background and foreground regions. However, the determination of a proper threshold value is typically dependent on subjective assumptions or empirical rules. In the chapter "Image Thresholding Based on Fuzzy Particle Swarm Optimization", the authors propose and analyze an image thresholding technique based on a fuzzy particle swarm optimization for efficient image segmentation.

Electrical Impedance Tomography (EIT) is a non-invasive imaging technique free of ionizing radiation. EIT image reconstruction is considered an ill-posed problem and, therefore, its results are dependent on the dynamics and constraints of reconstruction algorithms. The use of evolutionary and bio-inspired techniques to reconstruct EIT images has been taking place in the reconstruction algorithm area with promising qualitative results. In the chapter "Hybrid Metaheuristics Applied to Image Reconstruction for an Electrical Impedance Tomography Prototype", the authors discuss the implementation of evolutionary and bio-inspired algorithms and its hybridizations to EIT image reconstruction.

The editor has tried to bring together some notable contributions in the field of computational intelligence involving hybrid metaheuristic techniques for the purpose of image analysis. These contributions will surely open up more research avenues in this direction given the fact that faithful image analysis still remains a challenging thoroughfare in the computer vision research community. This book will serve graduate students and researchers in computer science, electronics communication engineering, electrical engineering, and information technology as a reference book and as an advanced textbook for some parts of the curriculum. Last but not least, the editor would like to take this opportunity to extend heartfelt thanks to Mr. Ronan Nugent, Senior Editor, Springer, for his valuable guidance and constructive support during the tenure of the book project.

Kolkata, India   Siddhartha Bhattacharyya
January 2018

Contents

Current and Future Trends in Segmenting Satellite Images Using Hybrid and Dynamic Genetic Algorithms

Mohamad M. Awad

Abstract Metaheuristic algorithms are an upper level type of heuristic algorithm. They are known for their efficiency in solving many difficult nondeterministic polynomial (NP) problems such as timetable scheduling, the traveling salesmen, telecommunications, geosciences, and many other scientific, economic, and social problems. There are many metaheuristic algorithms, but the most important one is the Genetic Algorithm (GA). What makes GA an exceptional algorithm is the ability to adapt to the problem to find the most suitable solution—that is, the global optimal solution. Adaptability of GA is the result of the population consisting of "chromosomes" which are replaced with a new one using genetics stimulated operators of crossover (reproduction), and mutation. The performance of the algorithm can be enhanced if hybridized with heuristic algorithms. These heuristics are sometimes needed to slow the convergence of GA toward the local optimal solution that can occur with some problems, and to help in obtaining the global optimal solution. GA is known to be very slow compared to other known optimization algorithms such as Simulated Annealing (SA). This speed will further decrease when GA is hybridized (HyGA). To overcome this issue, it is important to change the structure of the chromosomes and the population. In general, this is done by creating variable length chromosomes. This type of structure is called a Hybrid Dynamic Genetic Algorithm (HyDyGA). In this chapter, GA is covered in detail, including hybridization using the Hill-Climbing Algorithm. The improvements to GA are used to solve a very complex NP problem, which is image segmentation. Using multicomponent images increases the complexity of the segmentation task and puts more burden on GA performance. The efficiency of HyGA and HyDyGA in the segmentation process of multicomponent images is proved using collected field samples; it can reach more than 97%. In addition, the reliability and the robustness of the new algorithms are proved using different analysis methods.

M. M. Awad (✉)
National Council for Scientific Research, Remote Sensing Center, Beirut, Lebanon
e-mail: mawad@cnrs.edu.lb

© Springer International Publishing AG, part of Springer Nature 2018
S. Bhattacharyya (ed.), *Hybrid Metaheuristics for Image Analysis*,
https://doi.org/10.1007/978-3-319-77625-5_1

Keywords Dynamic · Genetic algorithm · Land cover · Metaheuristic · Segmentation

1 Introduction

Currently, the world is witnessing fast changes which include the invention of new technologies to increase the benefits for humanity. These new technologies require accurate, reliable, and instantaneous solutions for any sudden complex nondeterministic polynomial (NP)-complete problems. The solution most of the time is not a complete and definite one such as in the traveling salesman problem, timetable scheduling, and image segmentation. So, efficient algorithms are required to find solutions for these problems and that can provide the global optimum solution. Heuristic and metaheuristic algorithms can solve these problems. This chapter covers the metaheuristic algorithms, specifically the Genetic Algorithm (GA) for the segmentation of more complex types of images, namely remote sensing multicomponent satellite images.

1.1 Heuristic and Metaheuristic Algorithms

Heuristics is a word originated from the Greek language. It means discovering or finding. Many complex problems are known to be solved by the heuristics method which makes an important candidate for solving difficult problems. The common tactic between these disciplines is to search for a number of possibilities to find the best solution by selecting and accepting some of the found solutions. Thus, a heuristic is an algorithm that can explore all possible states of the problem, or that can discover the best ones. As indicated before, purely heuristics-based solutions may be unpredictable and often biased, and most of the time they lead to the local optimal solution. To avoid these problems, metaheuristics are presented [1]. Meta is a Greek word which means beyond. Thus, metaheuristics is thought of as an upper level heuristics, but in general it performs better than simple heuristics. The word metaheuristic was created by Glover [1]. All modern nature-inspired algorithms are known as metaheuristics [1–4], such as the Genetic Algorithm (GA), Scatter Search, Simulated Annealing (SA), Tabu Search, Ant Colony Optimization, Particle Swarm Optimization, the Differential Evolution, Firefly Algorithm, and the Bee Algorithm [1]. Metaheuristics consist of two processes, exploitation and exploration [1, 3]. In exploitation, the algorithm generates varied solutions to explore the search space globally, while in exploration the algorithm emphasizes the search in a local region to find an acceptable solution. A good balance between exploitation and exploration should be found by selecting the best solutions to improve algorithm convergence. Mixing these two processes ensures that a global optimum can be obtained. Metaheuristic algorithms are nature inspired and deploy either a population of solutions or one solution to explore the search space. Holland

[5] invented the GA in the early 1970s, a decade later, the growth of metaheuristic algorithms reached its highest peak. The next important step was the creation of SA in the early 1980s by Kirkpatrick et al. [6] which was motivated by the annealing process of metals. Glover [2] introduced the use of memory in a metaheuristic algorithm called Tabu Search. The process records the search and places them in a Tabu file in order to avoid revisiting previous solutions. Koza [7] introduced a new metaheuristic called Genetic Programming that created the structure of a whole new area of machine learning. In 2008 a Biogeography-based Optimization Algorithm inspired by biogeography was proposed by Simon [8]. This researches into the distribution of biological classes over time and space. Meanwhile, Yang [9] proposed a Firefly Algorithm, and later in 2009 Yang and Deb [10] developed a Cuckoo Search Algorithm. In 2010, Yang [11] also proposed a Bat Algorithm.

1.2 Image Segmentation

Segmentation divides an image into clusters with homogeneous attributes [12]. It is a significant attempt at successful image analysis. Segmentation is a difficult part in image processing because it controls the quality of extracted information that is analyzed [13]. There are many image segmentation methods which include edge detection [14–17], artificial neural networks (ANNs) [18, 19], and region growing [20, 21]. However, the best known ones are the clustering methods which are unsupervised with a random selection nature that is established on statistics. Formally, clustering splits an image I into k non-overlapping subsets

$$C = C_1, C_2, \ldots, C_k$$

such that

$$C_1 \bigcup C_2 \bigcup \ldots \bigcup C_k = I$$

and

$$C_i \bigcap C_j = \emptyset$$

$$i \neq j$$

There are many clustering methods such as Iterative Self-Organizing Data [22] and Fuzzy C-Means [23]. Much of the literature has applied GA to the image segmentation task, such as [24–28]. Finally, object-oriented segmentation methods are more precise in handling high resolution satellite images; however user intervention is required to provide more information to guide the method towards the solution. The software eCognitionTM by Definiens [29] is an example of an object oriented application for image segmentation. This type of software must be supported by pre-required information, such as contextual or textual. This is necessary to increase the segmentation result accuracies and to make them more appropriate for use [30]. In turn this software is not efficient for the different types of multicomponent images. Image segmentation methods can be either parametric or nonparametric. Several reasons make the parametric statistics unfavorable for some image segmentation methods. These can be: (1) complexity and inconsistency of the data and image design which depends on specific technology; (2) defectiveness in the process of image acquisition such as inherent noise and uncontrolled resolution that limits its efficiency; (3) different correction requirements such as enhancement, filtering, and the fusion process. To fix these limitations is time consuming, therefore leading to exhaustion of computer and human resources. In response to the need for effectiveness in statistical analysis, nonparametric segmentation methods have been widely used in solving many problems. These methods estimate data distributions without any assumptions about the structure of these data. Nonparametric methods are deployed when the problem of parameterization is unavailable. GA is a well-known nonparametric algorithm which is used widely in solving many problems including image segmentation. Although GA solves the problems encountered in parametric segmentation methods, it still needs to be enhanced to satisfy some critical issues for some applications. These issues are high processing speed and at the same time obtaining a global optimal solution. That is why Hybrid GA is used which includes heuristic algorithms such as Hill-Climbing that plays a role in slowing the convergence process toward a local optimal solution and providing the desired global optimal solution. Hybrid GA provides the desired solution but it is very slow and needs a long time to converge. To increase the speed of Hybrid GA, variable length chromosomes are used to create a dynamic population which reduces the time to perform the reproduction processes and increase the speed of Hybrid GA convergence toward the global solution significantly. Image segmentation techniques and methods can be divided into three supervised categories which require complete interference by the user and training. Semi-supervised categories require partial interference such as providing the number of regions, clusters, or classes. Finally, the unsupervised methods which do not require any interference by the user are completely automated. Some of the real-world applications of image segmentation are machine vision, medical imaging, biometrics, natural resources mapping, and object recognition.

1.3   Characteristics of the Remote Sensing Images

Remote sensing consists of many elements such as sensors, data, and procedures which are crucial for extracting information about the characteristics of the Earth's surface (i.e. the land, atmosphere, and oceans) which can be achieved with no direct physical contact. Information is obtained by capturing the reflected electromagnetic wave from the Earth's surface and its difference as a function of wavelength, angle direction, phase, location, and time. A variety of sensors are generally deployed both passively (i.e. which depends on reflected solar radiation or emitted radiation) and actively (i.e. which generates its own source of electromagnetic radiation). These sensors operate all over the electromagnetic spectrum from visible to microwave wavelengths. There are different platforms on which these sensors are attached, such as Earth-orbiting satellites, aircraft (most commonly), helicopters, and also balloons. Normally, remote sensing images are categorized by four different characteristics based on the type of sensor technology. The first characteristic is the spatial resolution where the higher the resolution the more detail about objects on Earth can appear in the image (the highest resolution is a few centimeters, while the lowest can reach 9 km). The lower the resolution the more area is captured. The second characteristic of the remote sensing images is the spectral resolution. It defines the capacity of the sensors to differentiate between wavelengths in the electromagnetic (EM) spectrum (bands). The finer the resolution the more images (bands) are collected which means that it is possible to obtain more detail about the reaction of sensed objects to the light. There are two technologies, hyperspectral and multispectral, where the first can have a spectral resolution of 1 nm with hundreds of bands compared to coarser spectral resolution, and the last one that can reach 100 nm with few bands. The third characteristic is the temporal resolution which defines the period of time needed to revisit the same spot on Earth. This is only applicable to satellite remote sensing. The less time needed to revisit a specific area the higher is the temporal resolution. The fourth and last characteristic is the radiometric characteristic. The radiometric resolution specifies how well the differences in brightness in an image can be perceived; this is measured through the number of levels of the gray value. The maximum value related to radiometric resolution is defined by the number of bits (binary numbers). An eight-bit representation has 256 gray values, a 16-bit representation has 65,536 gray values. The finer or the higher the radiometric resolution is the better are the recorded reflected waves, though the volume of measured data will be larger.

1.4 Satellite Image Types and Sources

Satellite images are sometime captured by sensors carried by a vehicle that orbits the Earth at a specific speed and a specific height—a satellite. Normally, the orbit can be at any height that ranges between 400 and 36,000 km. There are two types of satellite image: geostationary and polar, where the first covers a large fixed area and is taken from the remarkable height of 36,000 km. An example of these satellites is the famous Geostationary Operational Environmental Satellites (GOES) [31]. These satellites can capture an area with a swath width of 7000 km and with a spatial resolution lower than 4 km. The GOES images are used mainly to monitor weather and climatic conditions and natural disasters such as tornadoes. The other type of satellites are polar, which orbit at an elevation that ranges between 400 and 1000 km. These satellites have different characteristics which make them more attractive for use in different scientific and military tasks. There are many examples of these satellites such as the free of charge Landsat 7 and 8 which are still operational with a temporal resolution that can reach 8 days using both satellites. These medium to high spatial resolution satellite images can be downloaded from the USGS site [32]. Other free-of-charge satellites are Modis Aqua and Terra [33], both providing an image with a spatial resolution that ranges between 250 and 1000 m and which can be captured in a temporal resolution of less than 1 day. There are many other commercial satellites which are characterized by having a high to very high resolution which ranges from 31 cm such as Worldview-4 [34] to 5 m such as GeoEye [35]. There are more of these commercial satellites such as the 2.5 m SPOT 5, the 1.5 m SPOT 6 and 7, and the 0.5 m for Pleiades-1A and Pleiades-1B satellites [35].

2 Evolutionary Algorithms

Evolutionary algorithms are used for solving NP problems and to obtain global optimal solutions. They are stimulated by biological evolution and are considered a part of soft computing for those interested in studying these algorithms. Technically, they are considered to be a group of problem solvers with metaheuristic charac- teristics. These algorithms include methods stimulated by natural evolution such as reproduction and mutation. The solution of an optimization problem is selected from a group of individuals in the population, and the cost function determines the location of the solution. Any evolutionary algorithm includes three main powers that form the foundation of the optimization systems: selection, recombination, and mutation , which create the essential variety and thereby simplify innovation, although some consider selection as the major reason for increasing the quality of the global optimal solution. In this chapter GA is covered with the objective to use it in the processing of satellite images. In addition, several other heuristic and exhaustive algorithms are combined with GA to improve the segmentation process.

2.1 *Genetic Algorithm*

GAs were introduced by Holland in the early 1970s. Holland's unique goal was not to design algorithms to solve specific problems, but to study the concept of adaptation as it occurs in nature. He was looking to develop methods in which natural adaptation can be introduced into computer systems. Adaptation in Natural and Artificial Systems [5] by Holland introduced the GA as a generalization of evolution and provided a theoretical framework for adaptation under it. Holland's GA is a method for moving from one population to a new population by using what is called natural selection together with the genetic operators. The chromosome consists of genes such that each gene is an instance of a particular allele with a value of 0 or 1. The selection operator picks the chromosomes in the population in order to reproduce; in general the best chromosomes reproduce more than the less fit ones. Mutation randomly changes the allele values of random selected locations in the chromosome. The chromosomes in a GA population typically take the form of bit strings or integer numbers. Each gene in the chromosome has two possible alleles: 0 and 1 for bit strings or any value Z. In the case of images it is an integer value that ranges between 0 and 2^{Rbits} where Rbits is the radiometric resolution. The GA processes populations of chromosomes, successively replacing one such population with another. The GA requires a fitness function that scores (fitness value) to each chromosome in the current population. The fitness of a chromosome depends on how well that individual is able to solve the problem at hand. The following pseudo-code shows how GA works:

```
Randomly generate a population
Start with generation 1
Compute fitness
Select the best parent having the best fitness
While termination conditions are not met
Select randomly two parents for reproduction
Replace parents with the new off springs
Mutate chromosomes randomly
Compute fitness
Select the best parent having the best fitness
Replace old best individuals with the new best one
Increment generation
End while
```

Pseudo-code of Genetic Algorithm

2.2 Hill-Climbing Algorithm

Hill-Climbing is another optimization method for handling many NP problems. In this method, a local optimal solution can be started and the solution is improved repeatedly until some condition is maximized [36]. The idea of starting with a local optimal solution is matched to start from the base of the hill, improving the solution is compared to climbing a hill, and finally maximizing some conditions is compared to reaching the top of the hill. Hill-Climbing looks only at the current state and immediate future state. Hence, this technique is memory efficient as it does not maintain a search tree. In this iterative progressive method, the solution is achieved by making improvements towards an optimal choice in every iteration. However, this technique may encounter local maxima. In this situation, there is no nearby state for a better choice. In general the Hill-Climbing Algorithm is used in combination with the GA to slow GA termination or convergence toward a local optimal solution. In this case Hill-Climbing moves the local optima trapped in a valley toward a hill and then provides the solution to GA to continue searching for the global optimal solution. The following pseudo-code depicts how the Hill-Climbing Algorithm works.

```
Function Hill-Climbing (specific_problem)
returns a result which is the local optimal solution
Recent: best separable fitness value
Do loop
Neighbor: a highest-value replacement of recent one
if neighbor Less than or equal Recent then return recent one
Else
Recent=Neighbor
End If
End Do
```

Pseudo-code of Hill-Climbing Algorithm

The Hill-Climbing Algorithm is the most simple local search method. At each run the current individual is replaced by the best neighbor. In this case, it indicates that the best neighbor is the one with the highest fitness value. In more detail, if a heuristic cost estimate Cos is used, it is possible find the neighbor that has the lowest Cos.

2.3   Hybrid Genetic Algorithm

Generally, a simple GA comprises of several operations such as: (1) selection; (2) crossover and mutation; and (3) replacement. Genetic operations such as crossover are where two chromosomes are selected to reproduce new children, and mutation is the process of changing randomly one gene from its existing type to another different one. Finally, replacement is the task of substituting two parents with the new children. A Hybrid GA (HyGA) is a GA with a Hill-Climbing Algorithm which tests neighboring points in the search space and improves the fitness of chromosomes. It is a method capable of finding local extrema, as mentioned in [37]. HyGA starts as a first step by generating an unsystematic first population $P(0) = p_1, p_2, \ldots, p_n$. The population is created from random genes or obtained from another method as will be shown in the following subsections. Figure 1 shows how the Hill-Climbing Algorithm works with GA to achieve a global optimal solution.

Fig. 1 Hybrid genetic algorithm

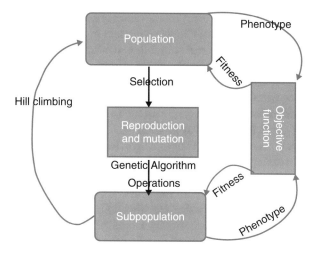

2.4 Example of HyGA Segmentation

The first task in HyGA is to create an initial population $P(0)$. These chromosomes are created using cluster centers which are obtained randomly from a satellite image I_S and by allocating image pixels to these selected clusters. These cluster centers are represented by the gray level values (value of pixels) in the image with multi-bands. Normally, a chromosome represents an image I_S (or portion of an image $I_P \in I_S$) such that four alleles represent multi-values of the gray level in three different EM spectra for a pixel and the cluster center. The objective function is depicted in Eq. (1) and is applied to calculate the closest pixels to the allocated center.

$$\text{Min}\left(\sum_{j=1}^{f}\sum_{i=1}^{y}\left[\text{Value}\left(Cce_j\right) - \sum_{r=1}^{3}\text{Value}(\text{pixel}_{ir})\right]\right) \qquad (1)$$

where f is the number of clusters in one individual, and Value (Cce_j) is the multi-band values of the center Cce_j. The final weights are summed and then they are multiplied by 255. Value (pixel$_{ir}$) are the multi-band values of the pixel assigned to the cluster center Cce_j in the chromosome ($\ldots$ pixel$_{i,1}$ pixel$_{i,2}$ pixel$_{i,3}$ Cce_j pixel$_{i+1,1}$ pixel$_{i+1,2}$ pixel$_{i+1,3}$ Cce_j $\ldots$), and y describes how many pixels exist in each cluster. The fitness F(Chromosome) is the objective function. Based on the fitness value, a number of chances are given to each chromosome to be selected using the roulette wheel process [38], where the lower in cost is given more chances for reproduction. Two chromosomes are selected randomly, and they are mated to create two new children. A mutation operator handles individual chromosomes by changing a cluster value with a different randomly selected one from the existing clusters in the image. The probability of crossover can vary between 40% and 70%, while mutation can vary between 10% and 20%. These reproduced children replace their parents, and their fitness values are calculated again. The reproduction process is tailed by the Hill-Climbing process in order to slow the fast convergence of GA toward a local optimal solution. In the first experiment, SPOT 5 is used where three bands are enhanced with respect to spatial resolution using the panchromatic image. The size of the SPOT image is 360×360 pixels and the resolution is 5 m; see Fig. 2a. The SPOT remote-sensing satellite program was created in France in partnership with other European countries such as Belgium. The deployment of SPOT satellites in orbit facilitated the mission of observing basically the entire planet in 1 day. Field work is carried out to verify the classified image where 110 different samples are collected from three classes (1—urban settlements (brown), 2—bare soil (light green), and 3—vegetation (green)). The results are verified based on a collection of samples combined with the confusion matrix [39]. The matrix covers information about real and projected classifications done by a specific method. The efficiency of such a method is normally estimated using the data in the matrix. The confusion matrix (Table 1) demonstrates that the accuracy value of the HyGA method is 82%.

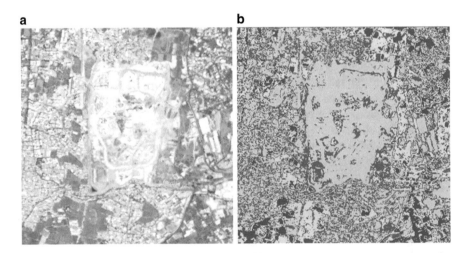

a   **b**

Fig. 2 Spot 5 image. (**a**) Original. (**b**) Segmented using HyGyA

Table 1 HyGA matrix of the processed SPOT 5 image

Classes	1	2	3	Total
1. Urban settlements	34	3	2	39
2. Bare soil	4	32	1	37
3. Vegetation	3	1	30	34
Total	41	36	33	110

3 Dynamic Genetic Algorithm

Dynamic means changeable, adaptable, and flexible with respect to changes in requirements of the problem to solve. Here Dynamic GA (DyGA) is related to changing population structure to create reliability and efficiency in solving a problem. Normally, the population of the standard GA is static, in other words there is no possibility to change the chromosome length. This is considered as a problem with respect to the speed of GA and the possibility to obtain a global optimal solution. In DyGA the structure of the chromosomes is different (the number of genes is variable) and the population can change in size based on the problem requirements.

3.1 Structure of the Dynamic Genetic Algorithm

DyGA is different from other conventional evolutionary algorithms, specifically traditional GA. The difference comes from its capacity to use a variable length of chromosomes with a different number of genes, such that each chromosome terminates with a symbol to mark the completion of the chromosome length (Fig. 3). Two chromosomes are chosen randomly to reproduce and two new children (siblings) are created. DyGA success to reach a global optimal solution depends on many factors; one of them is the probability of operators such as crossover. This should be carefully selected to avoid fast convergence to a local optimal solution which can be done by running many tests. The two new obtained children substitute for their parents and the fitness is computed for each individual. There are two main conditions that must be satisfied for the successful application of any new dynamic structure of a chromosome: (1) The mutation operator should be applied only on the genes, while symbols should be avoided; (2) Care should be taken in selecting the location of the crossover and that the mutation points away from symbols. DyGA use has many advantages such as minimizing the required time

Fig. 3 Dynamic population

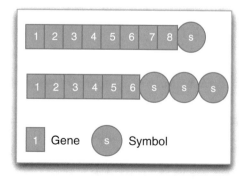

to find an optimal solution. The cause of this improvement is due to the new structure of the chromosome. The length is variable in terms of the number of genes allocated to each chromosome. Moreover, the time spent in executing operators for reproduction purposes and in computation (objective and fitness functions) is much less than that required by a normal GA with a static chromosome. The dynamic length chromosome guarantees variability in the population, which can easily provide the global optimal solution within the minimum time requirement. DyGA can be hybridized; in this case it is called Hybrid DyGA. In addition, the Hill-Climbing method is deployed to find adjacent points in the search space, and to help in directing the solution to the best fitness value. The Hill-Climbing process inspects cluster centers in each chromosome in detail and frequently modifies the chromosome to increase its fitness value. Hill-Climbing is an exploitation technique capable of finding local minima [37]. To understand the new dynamic method, a pseudo-code of HyDyGA is presented below. It is noticeable that HyDyGA resembles HyGA; however, the reproduction process is different with respect to the selection points for exchanging genes between chromosomes and with respect to mutation.

```
Begin
Read an image (e.g. satellite)
Begin HyDyGA
gen:=0 {count number of generations }
Create initial population Pop(gen)
Chromosomes have variable lengths
Evaluate population Pop(gen) While not completed (additional cycles) do
Hill-Climbing
gen= gen+ 1
Select two chromosomes I(gen) from Pop(gen)
Choose correct positions of the genes not symbol
Run Crossover Pop(gen)
Run Mutate Pop(gen) based on fitness value
Assess Pop(gen)
End while
End HyDyGA
Validate the results
```

3.2   Example of Hybrid Dynamic GA (HyDyGA)

In this example a multispectral image is segmented using HyDyGA. The Landsat image with spatial resolution of 30 m (Fig. 4a) is first read then the population of n chromosomes is created such that $Pop(0) = \{Ch_1, Ch_2, \ldots, Ch_n\}$. Each chromosome

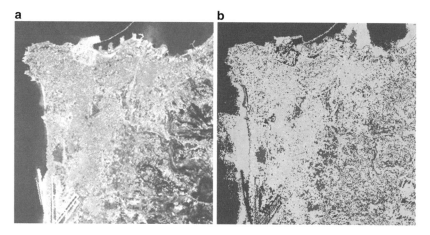

Fig. 4 (**a**) Original image Landsat 7. (**b**) HyDyGA solution

Table 2 HyDyGA matrix for Landsat 7

Classes	1	2	3	Total
1. Urban settlements	30	1	0	31
2. Vegetation	1	27	3	31
3. Bare soil	1	3	34	38
Total	32	31	37	100

comprises randomly selected cluster center values Pc and the values (three bands) for the pixels px, which are selected from the image. The position of these pixels in the chromosome are normally on the left side of each cluster center ($px_{i,1}px_{i,2}$ $px_{i,3}$ Pc_j $px_{i+1,1}px_{i+1,2}$ $px_{i+1,3}Pc_{j+1}$). A maximum number of cluster centers are used in the creation of the initial population to guarantee adequate diversity between different chromosomes. The objectives are to minimize the function in Eq. (1) and to optimize the number of clusters. Each individual fitness Fit(Ch$_i$) in the current population is proportional to the objective function. Each chromosome is allowed a number of chances to be selected by the roulette wheel process based on the fitness value, where the higher the fitness value the greater are the chances for reproduction. In the reproduction process, random selection of two chromosomes is the first step, followed by mating to create children. The choice of crossover probability is based on several tests. The maximum and the minimum mutation rate are specified by the user based on experimental trial and error tests. The initial number of iterations in HyDyGA is 25, the crossover probability is 65%, and the mutation ratio is variable based on the value of an individual fitness. Figure 4b shows the result of Landsat image segmentation. The result is again verified using a collection of field samples and the confusion matrix [39]. The information about real and forecast classifications in the matrix helps in the assessment of the results. Performance of the classification method is based on the computation of specific statistics using the matrix data. Table 2 shows the confusion matrix for three classes.

Field investigations are completed by selecting samples from the segmented images. Each sample can be either one of the three different classes (100 for all classes) (1—urban settlements (green), 2—vegetation (blue), and 3—bare soil (brown)). The accuracy of the Landsat segmentation is 91% as it is calculated based on the confusion matrix.

4   New Methods of Cooperation Between Metaheuristics and Other Algorithms

Sometimes it is required to advance the performance of a specific algorithm to solve a certain problem by using metaheuristic algorithms. The complexity of the problem shapes how these programs cooperate in order to solve a specific problem. The objective is to make the algorithm adapt to the problem at hand in order to obtain the global optimum solution. It is thought that self-adaptation is improved if cooperation and competition between algorithms are involved. There are several types of cooperation between GA and other algorithms, such as fuzzy logic, ANN and Fuzzy C-Means. However, there are few papers which cover the segmentation of satellite images. One, a promising paper [40], segments Landsat 8 images using a semi-supervised method based on GA and trained using a Radial Basis Function Neural Network. In a research paper [41] the authors used multiple-kernel fuzzy C-means (MFCM), ANN, fuzzy logic, and GA to segment a satellite image to extract some urban features such as buildings and roads. According to the authors their method was able to extract roads with an accuracy close to 89%, compared to 80% for another which combines only MFCM and ANN. Another research paper [42] uses GA to optimize the weights for an ANN supervised Multi-Layer Percepteron (MLP) [43] algorithm, in order to extract clouds from a weather satellite image. The results of GA-MLP showed better accuracy compared to the results of the MLP algorithm. Fuzzy logic and GA cooperation has played an important role in advancing the satellite image segmentation process, but its use is still limited to adjusting the probabilities of the reproduction operators for GA during this segmentation process. Sumera and Turker [44] used fuzzy logic to adjust the probabilities of crossover and mutation during the segmentation process of high resolution images by GA. The method proved to be efficient with a kappa index that approached 0.88. In this section, unsupervised nonparametric metaheuristic algorithm cooperation with another two non-metaheuristic algorithms to segment satellite images is illustrated using two different examples. There are several reasons for selecting these two examples, such as the lack of papers which cover this area of research and to prove that these types of algorithms can solve many of the problems which were listed in the previous sections (accuracy and speed). This includes the cooperation of Hybrid Dynamic GA (HyDyGA) with Fuzzy C-Means (FCM) [45]; here the role of the metaheuristic process is to improve the performance of FCM in image segmentation. On the other hand, another process which is an ANN algorithm called Self-Organizing Maps (SOMs) [46] is used to provide the metaheuristic process GA with initial cluster centers to start from an advanced point in the space of satellite segmentation solutions.

4.1 Hybrid Genetic Algorithm (HyGA) and Self-Organizing Maps (SOMs)

Kohonen's SOMs is an unsupervised nonparametric Artificial Neural Network method (ANN), where SOMs transform patterns of random dimensionality into the responses of 2D arrays of neurons. An important characteristic of the SOMs is the capability to conserve the neighborhood relationships of the input pattern. A distinctive SOMs structure consists of an input and an output layer (Fig. 5). The number of input neurons is equal to the dimensions of the input data; the neurons are arranged in a 2D array where each input is completely connected to all units. The values of the initial weights are randomly created, and their influence on the final state decreases as the number of trials increases or decreases [47]. SOMs segmentation of a satellite image maps patterns from a 3D into a 2D space. The network size is determined by the image size which can be computed empirically. The network is depicted by a mesh of $n \times n$ neurons which represents cluster units, such that each neuron represents the values of a pixel in three different bands of the image. During the training phase, the cluster unit is elected as a winner based on an input pattern matching the unit weight. This matching is based on the minimum value obtained by using Euclidean distance (Eq. (2)).

$$\|v - W_l^{\lfloor k \rfloor}\| = \min_i \|v - W_i^{\lfloor k \rfloor}\| \qquad (2)$$

$Im = \bigcup_{i=1}^{n} R_i$ is the union of all regions covering the entire image Im. $R_i \bigcap R_j = \emptyset$ where v is the input vector, $W_l^{\lfloor k \rfloor}$ is the weight of the selected unit l at repetition k, and $W_i^{\lfloor k \rfloor}$ is the weight for neuron i at iteration k. This selected unit and a neighborhood around it are then updated. All the neurons within a certain neighborhood around the leader participate in the weight update process (Eq. (3)). This process can be described by an iterative procedure.

$$W_i^{\lfloor k+1 \rfloor} = W_i^{\lfloor k \rfloor} + H_{li}^{\lfloor k \rfloor}(x - W_i^{\lfloor k \rfloor}) \qquad (3)$$

Fig. 5 SOMs map

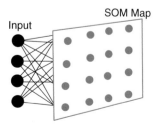

Input

SOM Map

where $H_{li}^{\lfloor k \rfloor}$ is a smoothing kernel which can be written in terms of the following Gaussian function:

$$H_{li}^{\lfloor k \rfloor} = \alpha^{\lfloor k \rfloor} \exp \left(-\frac{d^2\,(l,2)}{2\left(\mathrm{sd}^{\lfloor k \rfloor}\right)^2} \right) \qquad (4)$$

$H_{li}^{\lfloor k \rfloor} \to 0$ when $k \to T$ where T is the total number of iterations defined previously to be 1000, $\alpha^{\lfloor 0 \rfloor}$ is the initial guiding rate, and the default value is 0.1. The guiding rate is updated with every cycle as follows.

$$\alpha^{\lfloor k \rfloor} = \alpha^{\lfloor 0 \rfloor} \exp \left(-\frac{k}{T} \right) \qquad (5)$$

$\mathrm{sd}^{\lfloor k \rfloor}$ is the search distance at iteration k; initially, $\mathrm{sd}^{\lfloor 0 \rfloor}$ can be half the length of the network. As learning proceeds, the size of the neighborhood should be reduced until it includes only a single unit. The function is described by the following equation.

$$\mathrm{sd}^{\lfloor k \rfloor} = \mathrm{sd}^{\lfloor 0 \rfloor} \left(1 - \frac{k}{T} \right) \qquad (6)$$

After the SOMs reaches an unchangeable state, the image is transformed from an unlimited color space to a smaller dimension color space. In this space the number of colors is equal to the number of cells that form the network of SOMs. The final list of weight vectors after the network reaches stability are used as the new sample space. Each pixel's gray level value is represented by a neuron which stores the final weight. These results are used for clustering, by assigning from the weight values a set of cluster centers. The results obtained by the segmentation process of the image using SOMs is a local optimal one. It is expected that this solution is described as an over-segmented one where the following guidelines are not respected in the final result: i and j, there is no overlap of the regions. The violation of the above rules leads to over-segmentation. This is normally an NP problem; that is every time the same algorithm is run with different parameters (iteration numbers, network size) a different solution is obtained. To generate stability in the provided solution by SOMs it is essential to discover a global optimal solution. Usually, GA is an examining process which is built upon the laws of natural selection. Usually, it consists mainly of selection, genetic operations, and replacement. Genetic operations are crossover (reproduction) where two parents are selected to reproduce, and mutation is the process of altering one gene from one kind to another. Finally, replacement is the process of substituting two parents with the newly evolved children. An important characteristic of GA is its capacity to discover the global optimal solution without being stuck at the local minima [48]. Sometimes the complexity of the image segmentation problem makes it difficult to avoid falling into the local optima. To solve this matter a new procedure is added to GA; this new process is called Hill-Climbing. That is why the new technique is called Hybrid GA (HyGA).

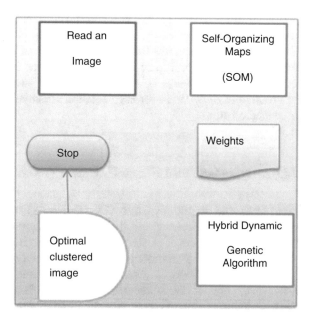

Fig. 6 SOMs-HyGA process

Hill-Climbing works by reducing the speed of convergence by penalizing individuals in the population (reducing fitness of the fittest). The segmentation procedure of the new method (Fig. 6) starts by reading a satellite image. Then SOMs uses the image features to combine the pixels into groups. The cluster center of each group is provided to HyGA to select the optimal solution in the segmentation process, by considering two criteria: (1) each group's total pixels, and (2) the proximity of pixels in the group to the cluster center. HyGA generates a population of individuals where the genes in each individual represent centers obtained from SOMs combined with the color value for each pixel in the multi-bands as part of the multispectral image. Equation (7) evaluates the fitness of chromosomes after each run, and the best solution is selected. Each chromosome consists of genes and each is a combination of the variable cluster center and the unchangeable pixel values obtained from the image. HyGA attempts to obtain the optimal number of classes (no over or under-segmentation). In other words, SOMs-HyGA fixes the problems encountered during the segmentation task because of deploying SOMs as a standalone segmentation algorithm. The objective function in Eq. (7) is used to compute the difference between pixels and the allocated centers. Another innovation of the new method is the ability to control the behavior of the (GA) by letting the Hill-Climbing process slow down the convergence and by controlling the elimination of individuals in the population.

$$\min \left(\sum_{j=1}^{kk} \sum_{i=1}^{z} \left[Va\left(P_j\right) - \sum_{r=1}^{3} Va(px_{ir}) \right] \right) \tag{7}$$

where kk is the number of the cluster centers in a chromosome, and $Va\,(P_j)$ is the multi-band values of the cluster center P_j. It is the sum of the obtained weights multiplied by 255. $Va(px_{ir})$ are the multi-band values of the pixel on the left side of the cluster center P_j in the chromosome. This method will lead HyGA to find an optimal number of classes (no under or over-segmentation). Moreover, SOMs-HyGA fixes the problem of under and over-segmentation which are normally the result of using one method alone.

4.2   Hybrid Dynamic (GA) and Fuzzy C-Means (FCM)

Fuzzy C-means (FCM) and hybrid dynamic GA (HyDyGA) can collaborate to improve the segmentation process. Typically, FCM clusters the image while HyDyGA finds the best arrangement of cluster centers that helps in minimizing the objective function of FCM (Eq. (8)) in order to obtain the global optimal solution. Given a set of n data patterns, $x = x_i, \ldots, x_n$, the FCM algorithm minimizes the weights within the group sum of the squared error objective function $J(U, V)$, where x_k is the kth p-dimensional data vector, v_i is the sample of the cluster center i, u_{ik} is the degree of membership of x_k in the ith cluster, and m is a weighting exponent on each fuzzy membership. The function $d_{ik}(x_k, v_i)$ is a distance measure between the data vector x_k and the cluster center v_i, n is the number of data vectors, and c is the number of clusters. A solution of the objective function $J(U, V)$ can be obtained via an iterative process where the degree of membership u_{ik} and the cluster center v_i are updated via Eqs. (9) and (10), respectively. The time is managed and allocated by HyDyGA while locating and then updating cluster centers. Another benefit gained by joining FCM and HyDyGA is the possibility to converge to a local optimal solution (which is a rare case) if the process fails. Then HyDyGA and FCM continues with the best cluster centers (local optimal) provided by the Hybrid Dynamic metaheuristic algorithm.

$$J\,(U, V) = \sum_{k=1}^{n} \sum_{i=1}^{c} u_{ik}^{m} d^{m}\,(x_k, v_i) \tag{8}$$

$$u_{ik} = \frac{1}{1 + \sum_{j=1}^{c} (d_{ik}/d_{ij})^{2/m-1}} \tag{9}$$

$$v_i = \frac{\sum_{k=1}^{n} u_{ik}^{m} x_k}{\sum_{k=1}^{n} u_{ik}^{m}} \tag{10}$$

subject to the following constraints:

$$u_{ik} \in [0, 1],\; \sum_{i=1}^{c} u_{ik} = 1\;\forall k,\; 0 < \sum_{k=1}^{n} u_{ik} < N\;\forall\, i$$

The alternative solution due to HyDyGA failure is to repeat the process with different individual chromosome structures and with different styles and probabilities of reproduction operators. The results are expected to be of different value, but HyDyGA can keep the process running with different parameters and structures until it locates a global optimal solution. For a better understanding of this new method, the pseudo-code provides a detailed and clear explanation of the steps involved in the whole process.

```
Begin the process
Read a satellite image
Start HyDyGA
gen: =0 initialization of generation counter
Initialize chromosomes Pop(gen)
Evaluate chromosomes Pop(gen) (i.e. compute fitness values)
While not over (more cycles) do
Randomize_Func (a heuristic process)
Hill-Climbing (Another heuristic)
gen:=gen+1
Select two chromosomes Pop(gen) from Pop(gen-1)
Run crossover Pop(gen)
Run Mutate Pop(gen) with fitness criteria
Assess Pop(gen)
End while
Provide the solution to FCM
Is it an optimal solution for FCM?
If no then change FCM parameters and continue with HyDyGA
Else (if yes)
End HyDyGA
Defuzzify and write final processed image
Evaluate the results
```

Pseudo-code of HyDyGA-FCM

The cooperative method works in series. First HyDyGA finds an optimal solution by running all the tasks required (initial population, selection reproduction, mutation). The results are then fed to FCM which in turn evaluates them and sends a response to HyDyGA. Normally, it is expected that the solution is the final one and that the result is unchangeable; then HyDyGA terminates and FCM creates the new segmented image. The method can be improved in the future by implementing the cooperative method as a parallel task where the communication between HyDyGA and FCM is done through sending signs as messages between different cooperating units or processors. In the case where no termination sign is sent by FCM, then HyDyGA keeps running, and the solutions are saved in a dynamic list. There are many advantages to these enhancements which include improvement of the efficiency of HyDyGA and increasing clustering speed.

4.3   Examples of Image Segmentation Using SOMs-HyGA

To demonstrate the practicality, strength, accuracy, and efficiency of the SOMs-HyGA an experiment on a Spot 4 satellite image is implemented (Fig. 7a), and at the end of the segmentation processes (Fig. 7b) a number of samples are collected from the segmented images which represent major classes. These samples are evaluated based on field work accompanied with advanced geospatial technologies such as a Global Positioning System (GPS). Later confusion matrices [39] are used to compute the accuracies. Each matrix consists of information about actual and predicted results which are created by the collection of field samples and the segmentation method. The performance of such systems is commonly evaluated using the data in the matrix. This task is completed with four classes (1 = crop_1 (light green), 2 = infrastructure/ urban (light brown), 3 = shrub (light pink), 4 = crop_2 green)) with hundreds of survey points (see Table 3). The accuracy of SOMs-HyGA is 92%.

The second experiment is implemented on a different type of satellite image, the IKONOS image (Fig. 8a) that has high spatial resolution of 1 m. The SOMs-HyGA is used to segment the IKONOS image with high quality (Fig. 8b) which improves

a   b

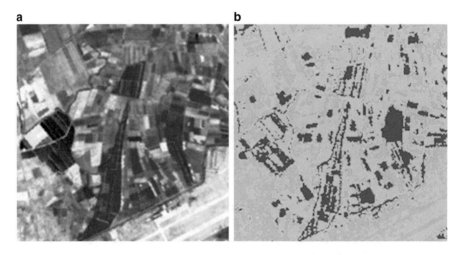

Fig. 7 Spot 4 satellite image. (**a**) Original. (**b**) Segmented by SOMs-HyGA

Table 3 Matrix of spot 4 image segmentation by SOMs-HyGA

Classes	1	2	3	4	Total
Crop_1	87	0	3	0	90
Infrastructure/urban	0	80	8	2	90
Shrub	5	0	85	0	90
Crop_2	0	10	0	80	90
Total	92	90	96	82	360

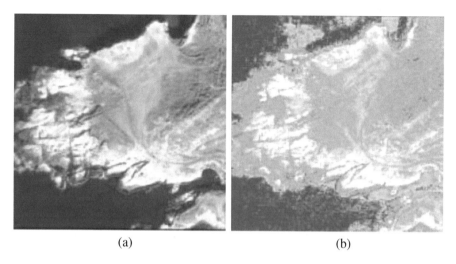

(a)   (b)

Fig. 8 IKONOS. (**a**) Original image. (**b**) Segmented by SOMs-HyGA

Table 4 Matrix of IKONOS image segmentation by SOMs-HyGA

Classes	1	2	3	4	Total
Water bodies	85	5	0	0	90
Vegetation	0	86	4	0	90
Bare soil	2	3	83	2	90
Rocks	0	0	2	88	90
Total	87	94	89	90	360

clearly this important step of image processing. Moreover, the confusion matrix is used to prove the high accuracy of the results (Table 4).

Four classes are used and evaluated with hundreds of samples collected in the field. These classes are the following: 1—water bodies (blue to dark blue), 2—vegetation (green), 3—bare land (gray), and 4—rock (white). The confusion matrix shows that the accuracy of SOM-HGA can reach 95%.

4.4 Examples of Satellite Image Segmentation Using FCM-HyDyGA

In this subsection, the FCM-HyDyGA cooperative method is used to segment two different satellite images: a medium spatial resolution image (30 m) and a high spatial resolution image (4 m). These two images are captured by Landsat ETM+ and IKONOS respectively. It should be noted here that both images consist of multibands, but only three spectral bands are used. In addition, IKONOS and Landsat satellite images are pan-sharpened with high to very high resolution panchromatic bands. The IKONOS image consists of many bands; one of them is the panchromatic

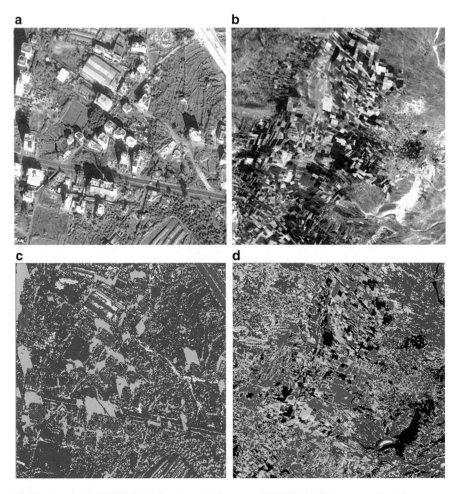

Fig. 9 Image. (**a**) IKONOS. (**b**) Landsat. (**c**) Segmented IKONOS. (**d**) Segmented Landsat

band. To improve the spatial resolution of the visible to near infrared (VNIR), these bands are fused "pan-sharpened" with panchromatic band (1 m) (Fig. 9a). In the case of Landsat the panchromatic fusion process increases the resolution to 15 m (Fig. 9b). HyDyGA is initiated with 30 iterations, but this can be increased without any dropping of the efficiency linked to the speed of convergence for the new method. Being a semi-supervised method FCM requires that the number of clusters be provided a priori. To overcome this obstacle, the new method runs HyDyGA first. This means that the optimal number of clusters and their values can be obtained before running FCM (optimal global solution). Segmentation of the images by FCM-HyDyGA resulted in different classes for each image such that in the case of the IKONOS image only four (Fig. 9c) classes are obtained while in the case of the Landsat image only six classes are obtained (Fig. 9d). Three important classes

Table 5 FCM-HyDyGA confusion matrix for Landsat 7 ETM+

Classes	1	2	3	Total
1. Urban settlements	23	2	0	25
2. Bare soil	0	22	0	22
3. Agriculture	0	0	27	27
Total	23	24	27	74

Table 6 FCM-HyDyGA matrix for IKONOS

Classes	1	2	3	4	Total
1. Vegetation_1	22	2	1	1	26
2. Urban settlements	0	21	0	3	24
3. Soil	0	2	29	0	31
4. Shadow	0	1	0	18	19
Total	22	26	30	22	100

in the segmented Landsat image are selected for evaluation (1—urban settlements (black), 2—bare soil (dark blue), 3—agriculture (dark green)), while all four classes are evaluated in the case of the IKONOS image (1—vegetation_1 (dark blue), 2—urban settlements (red), 3—soil (orange), 4—shadow (light blue)). The results of the segmentation method FCM-HyDyGA are evaluated using the confusion matrix [39] which consists of actual (field samples) and predicted (segmentation results) values. Tables 5 and 6 show the evaluation of both Landsat and IKONOS images. Normally, the distribution of the field samples plays an important role in the reliability of the accuracy. The distribution of these samples in this experiment ranges between uniform to random depending on the accessibility of the area of studies.

Using the confusion tables to evaluate the segmentation results of both Landsat and IKONOS by FCM-HyDyGA, one can determine that the accuracies of both segmented images to be 97% and 90% respectively. However, one can argue that the number of samples are not equal and at the same time not all the classes of Landsat images are included. But, considering all these factors we still believe that obtaining an accuracy equal to 90% for IKONOS and higher for Landsat is appropriate to prove the high efficiency and reliability of FCM-HyDyGA in providing a global optimal solution.

5   Metaheuristic Performance Analysis

In this section different analyses are conducted to show the efficiency and robustness of the metaheuristic algorithms. Landsat and IKONOS images of size 480×480 pixels (Fig. 9a, b) are used to analyze the speed and efficiency of these algorithms.

5.1 Metaheuristic Algorithm Complexity Analysis

The time complexity of SOMs is of the order $O(S*T*\text{Gen})$ where S and T are the size of Self-Organizing Maps Grid and Gen is the number of iterations. HyGA includes the Hill-Climbing process where the time complexity is equal to $\max(O(\text{Size(Pop)}, \text{Size(Ch)}), O(\text{Size(Pop)}*\text{Size(Ch)}*\text{Size(Gen)}))$, where Size(Ch) is the size of the chromosome, and Size(Pop) is the size of the population. The maximum rule states that SOMs-HyGA, which cooperates with both SOMs and HyGA, is the slowest with a time complexity equal to $\max(O(S*T*\text{Gen})$, $O(\text{Size(Pop)}*\text{Size(Ch)}*\text{Size(Gen)}))$. On the other hand, HyGA is the second segmentation method in speed followed by SOMs. FCM is more robust than the SOMs-HyGA method because FCM has a time complexity equal to $O(Cn^2)$, where C is the number of clusters and n is the data size. In FCM-HyDyGA, the estimation of cluster values is taken care of by the HyDyGA metaheuristic process which has a time complexity equal to $\max(O(\text{Size(Pop)}), O(\text{Size(Pop)}*\text{Size(Gen)}))$ which reduces the time complexity of FCM to $O(Cn)$. This is caused by having a dynamic number of cluster centers in short length chromosomes. That is why the Size(Ch) for HyDyGA is considered small and negligible (greater than or equal to 3 and less than the size of the image) compared to HyGA's Size(Ch) which is equal to the size of the image. This concludes that FCM-HyDyGA is the fastest with time complexity equal to $\max(O(Cn^2), O(\text{Size(Pop)}*\text{Size(Gen)}))$ followed in order by FCM, SOMs, and SOMs-HyGA.

5.2 Robustness and Efficiency Analysis

Several experiments are conducted with respect to speed and accuracy on the metaheuristic algorithm GA alone and in combination with other algorithms. The results of these experiments are compared to the results of SOMs and FCM

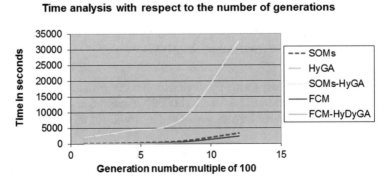

Fig. 10 Speed performance of metaheuristic methods

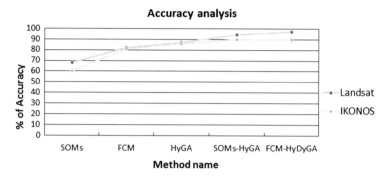

Fig. 11 Accuracy analysis of different metaheuristic methods

performances. Figure 10 shows the speed analysis for these algorithms. It is clear that HyDyGA performance is the best and is stable with the increase in the number of generations. Analyses are conducted to evaluate the accuracies of different segmentation methods including the metaheuristic ones and the results are displayed in Fig. 11. The graph shows that the highest accuracy is for FCM-HyDyGA followed by SOMs-HyGA, HyGA, FCM, and SOMs respectively. The provision of the initial solution by SOMs to HyGA makes their cooperation successful in providing a global optimal solution especially in the case of high resolution images which is comparable to the FCM-HyDyGA solution. However, the accuracy increases when FCM-HyDyGA is used to segment a mid-resolution image compared to SOMs-HyGA. The reasons are: (1) the Landsat image is more complex with respect to the variability of the features; (2) HyDyGA provides a global optimal solution to FCM as an initial solution.

5.3 Responsiveness Analysis

It is also important to analyze the reliability of metaheuristic algorithms with respect to responsiveness. This is done by adding more noise to the original image to be segmented. Noise adding should not reduce the accuracy, and the responsiveness of the segmentation method should be stable. To test this concept, the Landsat image in Fig. 9b is contaminated with different percentages of Gaussian noise [49], before being segmented with different segmentation methods. The responsiveness rate is calculated based on the following equation:

$$\varepsilon = 1 - \frac{A_{c0} - A_{ci}}{A_{c0}} \tag{11}$$

where ϵ is the responsiveness rate, A_{c0} is the accuracy of the segmented image without noise, and A_{ci} is the accuracy of the segmented image with $i\%$ of noise.

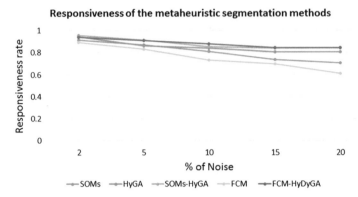

Fig. 12 Analysis of the responsiveness of metaheuristic methods to noise

Normally the responsiveness rate is between 0 and 1 where 1 is the best and 0 the worst. Figure 12 shows the responsiveness of the different methods with respect to noise adding. It is noticeable that all methods which depend on GA have the highest responsiveness rate. In addition, the responsiveness rates are different between different algorithms at low noise and stable for GA and cooperative algorithms. Analyzing the graph it is obvious that adding more noise drives SOMs and FCM toward the local optimal solution, and therefore the responsiveness is lower than GA.

6 Discussion

The segmentation process is known to be a very hard NP problem and the result of this process is very important for the success of the essential steps in image processing such as image classification, object detection, and object recognition. Many disciplines depend on the success of this important step in image processing. Example of these are the following: (1) traffic planning and control such as video surveillance [50], (2) land use planning such as mapping and natural resources management [51], (3) automation such as robotics and object extractions[52, 53], (4) biometrics such as face and finger print recognition [54], (5) medical image processing for the sake of locating tumors and virtual surgery simulation [55, 56], (6) in assessing the environment [57, 58], and (7) it is used in many other disciplines which lie outside the scope of this chapter. This large variety of use in different disciplines makes this process a critical one and forces the scientific community to look for more effective methods that can enhance the results of image segmentation than the existing ones such as supervised parametric methods. The main concern nowadays is to increase the accuracy of the results and the efficiency with respect to the speed of processing. Combining both objectives seems to be a difficult task especially with the appearance of more complex types of images such as multicomponent images where one of these images is the satellite image. This chapter has introduced several types of satellite images and illustrated the difference between them based on the spatial, spectral, temporal, and radiometric resolution. These differences complicate the task of image processing and add more burden on the segmentation methods. In this case the method must be adaptive to these changes and this is how it has been approached in this chapter where the metaheuristic GA has been modified to be Hybrid by including the Hill-Climbing method. This inclusion reduces the speed and slows down the process of image segmentation which can be noticed when increasing the satellite image size. To speed up the new modified GA a new term is introduced: the dynamic population or variable chromosome where in this new method the size of the chromosome is variable. In addition, part of the image is used in a random process of creation of the population which means that the required time for reproduction, replacement, and fitness calculation will be reduced significantly. Although the accuracy obtained can be judged as a good one, sometimes it is not higher than the one obtained using Hybrid GA alone. In addition, in the HyDyGA there are several issues which must be checked every time reproduction or mutation is performed, which is the difference in the size of the chromosomes and the position(s) of point(s) of selection in the crossover process. This is a very complex task which is performed successfully. The variability of the population is another issue that must be taken care of at the beginning when creating the population and in the evolution of a new population during the running of HyDyGA. One important topic which may improve the segmentation process, but may increase the complexity of the metaheuristic GA is the use of multi-objective GA. This topic is not tackled in this chapter due to the lack of sufficient experiments and work on this issue. However, it is worth listing some of these limited studies such as the ones in [59, 60]. These studies deal with image segmentation problems as problems having multiple objectives. This property can be defined as minimizing the distances between objects in the same cluster (intra-cluster), and maximizing the distances between different clusters (inter-cluster). Working with multiple objectives is considered a difficult problem, but sometime a multi-objective optimization approach for some problems is the only suitable method to find a solution [61]. Working with multi-objective GA adds more burden on the computer resources. It requires that the final result which is the best approximation of the Pareto front be considered as a multi-global optimum segmentation solution. On the other hand this problem is solved by using the combination of a metaheuristic algorithm such as GA and another clustering algorithm such as SOMs, which is the case with SOMs-HyGA or the combination of GA with FCM. In that case these processes can be run with different settings and can be evaluated to obtain the best global optimal solution.

7 Conclusion

Segmentation is one of the major steps in image processing without which object recognition would be obsolete. The segmentation problem is considered as a hard NP problem [62] and it cannot be solved with known conventional exhaustive algorithms and methods. Most of the known methods are either statistical parametric supervised, or statistical unsupervised . The lack of nonparametric unsupervised methods to segment different increasing types of complex multicomponent satellite images led the scientific community to think of metaheuristic algorithms as a solution for the segmentation of these types of multicomponent images. The success of these methods in solving many complex NP problems supported our choice of GA as a metaheuristic algorithm. GA is characterized by being efficient and robust in solving many different known NP problems. In this chapter, it has been shown that GA performance improves when another heuristic process such as Hill-Climbing is added. Moreover, changing the cluster size of centers in each chromosome increases the algorithm's efficiency. The suggested combination of Hill-Climbing with dynamic population improved the performance of GA with respect to finding the global optimal solution in less time compared to the conventional GA. The dynamic population of GA is created by changing the length of each chromosome and by adding a trivial number to complete every chromosome. This is done in order to ease and simplify the job of the crossover and mutation operators. Most of the time in normal GA, avoiding a local optimal solution requires further iterations as has been proved in much of the literature and in many experiments. However, sometimes the increase of iterations will lead GA to become stuck in the local optimal solution [63]. Here Hill-Climbing comes in handy to slow the convergence to a local optimal solution due to many factors such as changing the current solution provided by GA. This is done in order to prevent the metaheuristic algorithm from becoming stuck in the valley (local optimal solution) due to the Hill-Climbing process and moving it slowly toward the hill (global optimal solution). In addition, the experiments in this chapter covered the use of many different types of satellite images. It has been proved in this chapter that metaheuristic algorithms, specifically GA, can improve the segmentation process such that the accuracy of the segmentation can reach more than 97%. Finally it is suggested that Hybrid Dynamic GA can be further improved to solve more complex and large images by implementing a parallel version of the algorithm.

Acknowledgements The author thanks CNRS and the United States Geological Survey for providing satellite images which were used to prove many concepts in this chapter.

References

1. F. Glover, Future paths for integer programming and links to artificial intelligence. Comput. Oper. Res. **13**(5), 533–549 (1986)
2. F. Glover, M. Laguna, *Tabu Search* (Kluwer Academic, Norwell, 1997)
3. F. Glover, M. Laguna, R. Marti, Fundamentals of scatter search and path relinking. Control. Cybern. **39**(3), 653–684 (2000)
4. K. Dejong, *Evolutionary Computation a Unified Approach* (MIT Press, Cambridge, 2006), 268 pp.

5. J. Holland, *Adaption in Natural and Artificial Systems* (The University of Michigan Press, Ann Harbor, 1975)
6. S. Kirkpatrick, C.D. Gelatt, M.P. Vecchi, Optimization by simulated annealing. Science **220**(4598), 671–680 (1983)
7. J.R. Koza, *Genetic Programming: On the Programming of Computers by Means of Natural Selection* (MIT Press, Cambridge, 1992)
8. D. Simon, Biogeography-based optimization. IEEE Trans. Evol. Comput. **12**(6), 702–713 (2008)
9. X.S. Yang, Firefly algorithms for multimodal optimization, in *Stochastic Algorithms: Foundations and Applications*, ed. by O. Watanabe, T. Zeugmann. Lecture Notes in Computer Science, vol. 5792 (Springer, Berlin, 2009), pp. 169–178
10. X.S. Yang, S. Deb, Engineering optimization by cuckoo search. Int. J. Math. Model. Numer. Optim. **1**(4), 330–343 (2010)
11. X.S. Yang, A new metaheuristic bat-inspired algorithm, in *Nature Inspired Cooperative Strategies for Optimization (NICSO 2010)*. Studies in Computational Intelligence, vol. 284 (Springer, Berlin, 2010), pp. 65–74
12. R. Demirci, Rule-based automatic segmentation of color images. Int. J. Electron. Commun. **60**, 435–442 (2006)
13. W. Pratt, *Digital Image Processing*, 2nd edn. (Wiley, New York, 1991)
14. J. Canny, Computational approach to edge detection. IEEE Trans. Pattern Anal. Mach. Intell. **8**(6), 679–698 (1986)
15. J. Shen, S. Castan, An optimal linear operator for edge detection, in *Abstract of the Proceeding of IEEE International Conference on Computer Vision and Pattern Recognition* (1986)
16. M. Kass, A. Witkin, D. Terzopoulos, Snakes: active contour models. Int. J. Comput. Vis. **1**, 259–268 (1987)
17. R. Deriche, Using Canny's criteria to derive a recursively implemented optimal edge detector. Int. J. Comput. Vis. **1**(2), 167–187 (1987)
18. M. Awad, K. Chehdi, A. Nasri, Multi-component image segmentation using genetic algorithms and artificial neural network. IEEE Geosci. Remote Sens. Lett. **4**(4), 571–575 (2007)
19. M.M. Awad, An unsupervised artificial neural network method for satellite image segmentation. Int. Arab. J. Inf. Technol. **7**(2), 199–205 (2011)
20. F. Shih, S. Cheng, Automatic seeded region growing for color image segmentation. Image Vis. Comput. **23**(10), 877–886 (2005)
21. L. Garcia, E. Saber, S. Vantaram, V. Amuso, M. Shaw, R. Bhaskar, Automatic image segmentation by dynamic region growth and multiresolution merging. IEEE Trans. Image Process. **18**(10), 2275–2288 (2009)
22. J. Jensen, *An Introductory Digital Image Processing: A Remote Sensing Perspective* (Prentice-Hall, Upper Saddle River, 1996), 379 pp.
23. J. Bezdek, *Pattern Recognition with Fuzzy Objective Function Algorithms* (Plenum Press, New York, 1981)
24. G. Lo Bosco, A genetic algorithm for image segmentation, in *Proceedings of the 11th International Conference on Image Analysis and Processing (ICIAP01)*, Palermo, 2001
25. C. Lai, C. Chang, A hierarchical genetic algorithm based approach for image segmentation, in *Proceedings of the IEEE International Conference on Networking, Sensing and Control*, Taiwan (2004), pp. 1284–1288
26. V. Ramos, F. Muge, Image color segmentation by genetic algorithms, in *Proceedings of the 11th Portuguese Conference on Pattern Recognition 2000*, Porto (2000), pp. 125–129
27. S. Chabrier, C. Rosenberger, B. Emile, H. Laurent, Optimization based image segmentation by genetic algorithms. EURASIP J. Image Video Process. (2008). https://doi.org/10.1155/2008/842029
28. W. Fujun, J. Li, S. Liu, X. Zhao, D. Zhang, Y. Tian, An improved adaptive genetic algorithm for image segmentation and vision alignment used in microelectronic bonding. IEEE/ASME Trans. Mechatronics **19**(3), 916–923 (2014). ISSN 1083-4435

29. Ecognition, http://www.ecognition.com/category/related-tags/definiens-ecognition. Cited 3rd of July 2017
30. M. Marangoz, M. Oruc, G. Buyuksalih, Object-oriented image analysis and semantic network for extracting the roads and buildings from Ikonos pan-sharpened images, in *Proceedings of Geo-Imagery Bridging Continents, XXth ISPRS Congress*, Istanbul, 2004
31. National Oceanic and Atmospheric Administration (NOAA), GOES data archive, https://www.class.ngdc.noaa.gov/saa/products/search?datatype_family=GVAR_IMG. Cited 15 July 2017
32. United States Geological Survey Society (USGS), Landsat 8 and 7 download, https://eo1.usgs.gov. Cited 18 July 2017
33. National Aeronautical Space Agency (NASA), Modis Web site, https://modis.gsfc.nasa.gov. Cited 16 July 2017
34. Digital Globe, WorldView-4, http://worldview4.digitalglobe.com/#/main. Cited 16 July 2017
35. Satellite Imaging Corporation, http://www.satimagingcorp.com/satellite-sensors/other-satellite-sensors/. Cited 15 July 2017
36. S. Skiena, *The Algorithm Design Manual*, 2nd edn. (Springer Science and Business Media, Berlin, 2010). ISBN 1-849-96720-2
37. R. Huapt, S. Haupt, *Practical Genetic Algorithm* (Wiley, Hoboken, 2004)
38. J. Baker, Reducing bias and inefficiency in the selection algorithm, in *Proceedings of the 2nd International Conference on Genetic Algorithms and Their Applications* (1987), pp. 14–21
39. R. Kohavi, F. Provost, Glossary of terms, Special Issue Appl. Mach. Learn. Knowl. Disc. Process **30**(2/3), 271–274 (1998)
40. A. Singh, K. Singh, Satellite image classification using genetic algorithm trained radial basis function neural network, application to the detection of flooded areas. J. Vis. Commun. Image Represent. **42**, 173–182 (2017)
41. B. Ankayarkanni, A.E.S. Leni, GABC based neuro-fuzzy classifier with multi kernel segmentation for satellite image classification. Biomed. Res. J. (2016). ISSN 0970-938X. Special Issue: S158–S165
42. V. Preetha Pallavi, V. Vaithiyanathan, Combined artificial neural network and genetic algorithm for cloud classification. Int. J. Eng. Technol. **5**(2), 787–794 (2013)
43. N. Da Silva, D. Hernane Spatti, R. Andrade Flauzino, L.H.B. Liboni, S.F. dos Reis Alves, *Artificial Neural Networks* (Springer International Publishing, Berlin, 2017), 307 pp.
44. E. Sumera, M. Turker, An adaptive fuzzy-genetic algorithm approach for building detection using high-resolution satellite images. Comput. Environ. Urban. Syst. **39**, 48–62 (2013)
45. J. Bezdek, R. Ehlrich, W. Full, FCM: the fuzzy-C-means clustering algorithm. Comput. Geosci. **10**(23), 191–230 (1984)
46. T. Kohenen, *Self-organizing Maps*. Springer Series in Information Sciences, vol. 30 (Springer, Berlin, 2001), 501 pp.
47. H. Yin, N. Allinson, On the distribution and convergence of feature space in self-organizing maps. Neural Comput. **7**(6), 1178–1187 (1995)
48. S.C. Ng, S.H. Leung, C.Y. Chung, A. Luk, W.H. Lau, The genetic search approach. A new learning algorithm for adaptive IIR filtering. IEEE Signal Process. Mag. **13**(6), 38–46 (1996)
49. M. Jayasree, N.K. Narayanan, A novel fuzzy filter for mixed impulse Gaussian noise from color images, in *Proceedings of the International Conference on Signal, Networks, Computing, and Systems*, ed. by D. Lobiyal, D. Mohapatra, A. Nagar, M. Sahoo. Lecture Notes in Electrical Engineering, vol. 395 (Springer, New Delhi, 2017), pp. 53–59
50. D. Zhu, J. Jiang, The multi-objective image fast segmentation in complex traffic environment, in *International Conference on Mechanic Automation and Control Engineering (MACE)*, Wuhan, 2010
51. M. Awad, I. Jomaa, F. EL-Arab, Improved capability in stone pine forest mapping and management in Lebanon using hyperspectral CHRIS PROBA data relative to landsat ETM+. Photogramm. Eng. Remote Sens. **80**(5), 725–731 (2014)

52. N. Greggio, A. Bernardino, J. Santos-Victor, Image segmentation for robots: fast self-adapting Gaussian mixture model, in *International Conference Image Analysis and Recognition ICIAR 2010: Image Analysis and Recognition*. Lecture Notes in Computer Science Book Series, vol. 6111 (2010), pp. 105–116

53. M. Awad, A morphological model for extracting road networks from high-resolution satellite images. J. Eng. **2013**(2013), 9 pp. (2013). www.hindawi.com/journals/je/2013/243021/. Cited June 2017

54. X. Font-Aragones, M. Faundez-Zanuy, J. Mekyska, Thermal hand image segmentation for biometric recognition. IEEE Aerosp. Electron. Syst. Mag. **28**(6), 4–14 (2013)

55. M. Forghani, M. Forouzanfar, M. Teshnehlab, Parameter optimization of improved fuzzy c-means clustering algorithm for brain MR image segmentation. Eng. Appl. Artif. Intell. **23**(2), 160–168 (2010)

56. W. Wu, A. Chen, L. Zhao, J. Corso, Brain tumor detection and segmentation in a CRF framework with pixel-pairwise affinity and super pixel-level features. Int. J. Comput. Aided Radiol. Surg. **9**, 241–253 (2014)

57. N. Kabbara, J. Benkhelil, M. Awad, V. Barale, Monitoring water quality in the coastal area of Tripoli (Lebanon) using high-resolution satellite data. Elsevier ISPRS J. Photogramm. Remote Sens. **63**(5), 488–495 (2008)

58. M. Awad, Sea water chlorophyll-a estimation using hyperspectral images and supervised artificial neural network. Elsevier Ecol. Inform. **24**, 60–68 (2014). https://doi.org/doi:10.1016/j.ecoinf.2014.07.004

59. A. Mukhopadhyay, S. Bandyopadhyay, U. Maulik, Clustering using multi-objective genetic algorithm and its application to image segmentation, in *Proceedings of the IEEE International Conference on Systems - Man and Cybernetics*, vol. 3 (2007)

60. N. Ghoggali, F. Melgani, Y. Bazi, A multi-objective genetic SVM approach for classification problems with limited training samples. IEEE Trans. Geosci. Remote Sens. **47**, 1707–1718 (2009)

61. V. Guliashki, H. Toshev, C. Korsemov, Survey of evolutionary algorithms used in multi-objective optimization. Probl. Eng. Cybern. Robot. Bulgarian Acad. Sci. **60**, 42–54 (2009)

62. A. Sharma, S. Sehgal, Image segmentation using firefly algorithm, in *International Conference on Information Technology (InCITe) - The Next Generation IT Summit on the Theme - Internet of Things: Connect your Worlds*, Noida, 6–7 Oct. 2016. https://doi.org/10.1109/INCITE.2016.7857598

63. M. Rocha, J. Neves, Preventing premature convergence to local optima in genetic algorithms via random offspring generation, in *Multiple Approaches to Intelligent Systems*, ed. by I. Imam, Y. Kodratoff, A. El-Dessouki, M. Ali. Lecture Notes in Computer Science, vol. 1611 (Springer, Berlin, 1999)

A Hybrid Metaheuristic Algorithm Based on Quantum Genetic Computing for Image Segmentation

Safia Djemame and Mohamed Batouche

Abstract This chapter presents a new algorithm for edge detection based on the hybridization of quantum computing and metaheuristics. The main idea is the use of cellular automata (CA) as a complex system for image modeling, and quantum algorithms as a search strategy. CA is a grid of cells which cooperate in parallel and have local interaction with their neighbors using simple transition rules. The aim is to produce a global function and exhibit new structures. CA is used to find a subset of a large set of transition rules, which leads to the final result, in our case: edge detection. To tackle this difficult problem, the authors propose the use of a Quantum Genetic Algorithm (QGA) for training CA to carry out edge detection tasks. The efficiency and the enforceability of QGA are demonstrated by visual and quantitative results. A comparison is made with the Conventional Genetic Algorithm. The obtained results are encouraging.

Keywords Metaheuristics · Quantum computing · Quantum genetic algorithm · Complex systems · Image segmentation · Edge detection · Cellular automata · Rule optimization

1 Introduction

Quantum computing (QC) is an emerging interdisciplinary science that has induced intense research in the last decade. QC is based on the principles of quantum mechanics such as quantum bit representation and state superposition. QC is capable

S. Djemame (✉)
Computer Science Department, Faculty of Sciences, University of Ferhat ABBAS-Setif1, Setif, Algeria
e-mail: djemame@univ-setif.dz

M. Batouche
College of NTIC, University of Constantine 2, Constantine, Algeria
e-mail: mohamed.batouche@univ-constantine2.dz

© Springer International Publishing AG, part of Springer Nature 2018
S. Bhattacharyya (ed.), *Hybrid Metaheuristics for Image Analysis*,
https://doi.org/10.1007/978-3-319-77625-5_2

of processing a huge number of states simultaneously, so it brings a new philosophy to optimization due to its underlying concepts.

Metaheuristics has become the main interest for researchers in optimization. It has allowed the solving of many practical and academic optimization problems, and gives better results than classical methods. Though, the use of a unique metaheuristic is a bit restrictive, a smart hybridization of these concepts leads to more satisfying behavior and better results, especially when dealing with large scale problems. Combinations of metaheuristics with mathematical programming, machine learning, and quantum computing have provided very powerful search algorithms.

These approaches are commonly referred to as "hybrid metaheuristics". They reveal efficiency in solving the following complex optimization problems: continuous/discrete optimization, mono-objective/multi-objective optimization [18], optimization under uncertainty, combinatorial optimization [20], and classification [1]) in a diverse range of application domains.

This chapter deals with the hybridization of a metaheuristic: the Genetic Algorithm and the concept of quantum computing for solving an image processing problem (image segmentation).

The novelty of this work is twofold: firstly, a new approach is presented, combining the powerful principle of QC with a complex system CA, to deal with edge detection. To the best of our knowledge, especially in image processing, there is no prior work combining CA and QGA for solving edge detection problems.

Secondly, the use of QGA as a search strategy in the large space of transition rules gives an efficient solution to the hard problem of how to find the subset of rules which achieves the desired function.

The remainder of the chapter is organized as follows: In Sect. 2, related works are discussed. In Sect. 3, an overview of quantum computing is presented. In Sect. 4, the concepts inherent to the QGA are explained. In Sect. 5, the new proposed approach for edge detection using QGA and CA rules is illustrated. In Sect. 6, experimental results, both visual and numerical, are shown. In Sect. 7, a comparison is made between the QGA algorithm and the Conventional Genetic Algorithm (CGA), and experimental results are shown. Finally, conclusions and future perspectives are drawn in Sect. 8.

2   Related Works

In [17], the author proposed the first quantum algorithm for number factorization. In [4], the authors proposed a solution to the problem of random searches in databases by using a quantum algorithm. Researchers attempted to fit the features of QC into conventional algorithms. In the early 1990s, the hybridization of QC and evolutionary computation proved its efficiency when working on complex problems. Regarding this, QGA showed high ability for solving large scale optimization problems [6, 7]. QGA can handle the stability between exploration and exploitation more readily than classical genetic algorithms. A few individuals are sufficient for exploring the search space and finding a good solution within a short time [8]. QGA is also characterized by the representation of chromosomes, the search space, the fitness function, and the movements of populations. Besides mutation and crossover, a new operation called *interference*, introduced in [15], is also utilized. QGA is characterized by a small size of the population, a high speed of convergence, a great capability of global optimization, and good robustness.

QGA has been used for solving combinatorial problems [5] and function optimization [23]. In bioinformatics, QGA is used to solve the prediction of new protein sequences [13]. In the image processing domain, the area of interest, there are a few works in the literature which solve image processing problems like filtering and edge detection by using QGAs. In [24], the authors present a QGA for image segmentation based on maximum entropy. In [19], the authors present a quantum inspired genetic algorithm for multi-objective segmentation with a split/merge strategy. In [3], the authors applied quantum modeled clustering algorithms for image segmentation. Concerning the problem of automating rule generation, we found in the literature several works. For example, [16] used a Sequential Floating Forward Search Algorithm to achieve image processing tasks with CA. In [10], the authors used GA for evolving CA rules for edge detection. In [2] a quantum algorithm is used for image denoising.

In an attempt to make a contribution in this field, we propose the use of a QGA as a search strategy for finding CA rules from a very large space to perform edge detection on images. The algorithm proposed in this chapter benefits from QGA capabilities to explore wide search spaces and rapid convergence, and the power of CA to model images efficiently and extract a subset of rules which performs satisfactory edge detection.

3 Overview of Quantum Computing

During the 1980s, Richard Feynman, the American physicist, made influential contributions in the domain of quantum mechanics. He was also one of the first scientists to conceive the possibility of quantum computers. During the early 1990s, QC attracted increasing interest and gave rise to much research. The algorithmic complexity here is obviously reduced by the parallelism of QC. Such a capability can also be exploited to resolve optimization problems which need to be examined over large solution intervals.

3.1 Definition of a Quantum Bit

The smallest unit of information stocked on a quantum computer is called a quantum bit (qubit) [9]. A qubit lies within the superposition values of 0 and 1. The state of a

Fig. 1 Classical bit and
quantum bit [11]

Classical bit

Quantum bit

qubit is depicted by Dirac's equation:

$$|\psi = \alpha|0> +\beta|1> \tag{1}$$

α and β are complex numbers called the probability amplitudes of the corresponding
state of the qubit and satisfy the condition:

$$|\alpha|^2 + |\beta|^2 = 1 \tag{2}$$

$|0>$ is the classical bit value 0 and $|1>$ is the classical bit value 1. When we
measure the qubit's state, it gives 0 with a probability $|\alpha|^2$ and 1 with a probability
$|\beta|^2$ (Fig. 1).

3.2 Quantum Register

A quantum register is an arbitrary superposition of m qubits. In a classical system, a
register of m bits can represent only one value among 2^m possible values. However,
a quantum register can represent 2^m states simultaneously.

The size of the search space grows exponentially with the number of particles.
This suggests that instructions are executed with higher speed on quantum computers than on sequential computers. A quantum operation is performed in parallel on
all the states of the superposition[14].

3.3 Quantum Measure

The measure of a qubit state leads to either bit '1' or bit '0'. The result depends on
the values of the qubit's amplitudes. Figure 2 shows an example of qubit measure
which has a probability of 70% of being in state '1', and probability 30% of being
in state '0'.

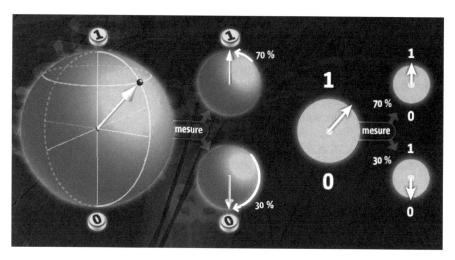

Fig. 2 Quantum measure [11]

3.4 Quantum Algorithms

A quantum algorithm is constituted by a succession of quantum operations, realized on a quantum system. They are executed sequentially using quantum gates and quantum circuits.

Quantum gates are elementary operations on a qubit or a quantum register. The most used quantum gates are NOT gates, controlled NOT gates, Hadamard gates, and rotation gates. The choice of the quantum gate depends on the problem to be solved

Quantum circuits are the combination of two or several quantum gates, allowing the application of a more complex processing on a quantum system.

Quantum algorithms offer a reduced algorithmic complexity, in comparison with classical algorithms, thanks to the superposition of states. Let us emphasize that the design of such algorithms is difficult. The design of a powerful quantum machine is prohibitively difficult, and developed quantum algorithms still need to be simulated on conventional processors [12].

4 Quantum Genetic Algorithm Principles

A QGA is a genetic algorithm where manipulated individuals are quantum chromosomes. Representation of the chromosomes rests on the principle of the qubit and is endowed with other quantum operations. This means that genetic operations are totally redefined to be adapted to the new representation of the chromosomes. Algorithm (1) shows the pseudo-code of QGA.

ALGORITHM 1
BEGIN
$Q(t = 0)$ is a population of qubit chromosomes at generation t
$P(t)$ is a set of binary solutions at generation t
Initially generate randomly initial population $Q(t = 0)$ of quantum chromosomes
Repeat until convergence
Generate $P(t)$ by measure of $Q(t)$
Evaluate $P(t)$
Save the best solution b
Update $Q(t)$ by quantum interference
$t \longleftarrow t + 1$
END.

4.1   Coding of Quantum Chromosomes

A quantum chromosome is a chain of n qubits, forming a quantum register. Table 1 illustrates the structure of a quantum chromosome.

4.2   Measuring Chromosomes

For exploiting effectively the superpositioned states in the qubit, we need to take a reading for each qubit. The purpose of this operation is the extraction of a binary chromosome from a quantum one. The purpose is to allow evaluation of the population's individuals according to the binary chromosomes extracted (Table 2).

The measure function can be easily represented by algorithm (2).

Let q be a qubit and 'measure' its measure function. q is depicted by:

$$|\psi = \alpha|0 > +\beta|1 > \tag{3}$$

Table 1 Structure of a quantum chromosome

α_0	α_1	α_2	α_3	$\ldots$	α_n
β_0	β_1	β_2	β_3	$\ldots$	β_n

Table 2 Measure of chromosomes

Quantum chromosome							
α_0	α_1	α_2	α_3	α_4	α_5	$\ldots$	α_n
β_0	β_1	β_2	β_3	β_4	β_5	$\ldots$	β_n

$\Downarrow$ *Measure*

0	1	1	0	0	1	$\ldots$	1
Binary chromosome							

ALGORITHM 2
BEGIN
r = get r randomly in [0, 1]
if $r > (\alpha)^2$ return 1
else return 0
END.

where:

$$|\alpha|^2 + |\beta|^2 = 1 \qquad (4)$$

The fitness value is obtained by the evaluation of each binary solution.

4.3 Quantum Genetic Operations

In this subsection, we describe the most important quantum genetic operations:

- Quantum crossover: this operation is similar to the crossover of a classical GA, except that it operates on the qubits of a quantum chromosome. The selected individuals will be randomly distributed in couples, then they begin to reproduce. The operation is realized by exchanging fragments situated after the cut points, which allows the creation of two new quantum chromosomes.
- Quantum mutation: the role of this operation is to change the value of some random positions in the quantum chromosome according to a mutation rate.
- Quantum interference: this operation allows the modification of individual amplitudes. The aim is the amelioration of their fitness. It consists in moving the state of every qubit in the direction of the value of the best found solution. This operation is useful to intensify the search around the best solution. It can be realized by the means of a transformation which allows a rotation. The angle of the rotation is a function of the amplitudes α, β and the value of the bit corresponding to the best solution. The value of rotation angle $\delta\theta$ must be chosen so as to avoid premature convergence.

5 The Proposed Approach

The problem we propose to solve is detecting edges in images. For modeling an image, we use a complex system: CA.

CA is a grid of cells which interact locally by simple rules and evolve towards a global complex behavior. Interactions between cells are defined with local rules. The set of all these rules forms the transition function of the CA.

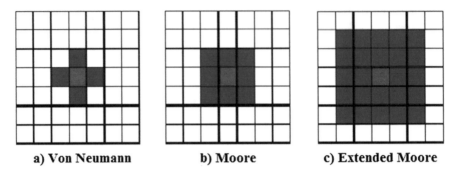

a) Von Neumann   **b) Moore**   **c) Extended Moore**

Fig. 3 Models of cellular automata neighborhood

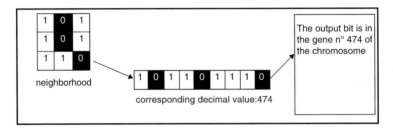

Fig. 4 Correspondence between CA and chromosome

In image processing, CA is used efficiently to model an image. A state of a cell is the color of the pixel. The transition rule is defined by the actual state of the cell and the state of the neighborhood. In this work, the neighborhood used is the one in "Moore" (Fig. 3).

The initial configuration of CA is the input image to be processed.

The final configuration is the output image (segmented, filtered).

The main problem with CA remains how to define the only subset of rules, among a big set, which gives rise to the desired effect. The number of transition functions grows with the number of cell states and the size of neighborhood. For example, for a Moore neighborhood and two cell states, the size of the transition function is 512, and the number of possible transition functions is 2^{512}.

5.1   From Cellular Automata to Chromosome

Each transition function is associated with a lookup table; its size is 512 positions. Each position corresponds to a specific configuration of the neighborhood. Figure 4 shows a possible way for corresponding a configuration of the neighborhood with its position in the lookup table, and then deducing the result value in the chromosome.

Table 3 Initialization of quantum chromosome

$\frac{1}{\sqrt{2}}$	$\frac{1}{\sqrt{2}}$	$\frac{1}{\sqrt{2}}$	$\frac{1}{\sqrt{2}}$	$\frac{1}{\sqrt{2}}$	$\frac{1}{\sqrt{2}}$	$\cdots$	$\frac{1}{\sqrt{2}}$
$\frac{1}{\sqrt{2}}$	$\frac{1}{\sqrt{2}}$	$\frac{1}{\sqrt{2}}$	$\frac{1}{\sqrt{2}}$	$\frac{1}{\sqrt{2}}$	$\frac{1}{\sqrt{2}}$	$\cdots$	$\frac{1}{\sqrt{2}}$

Table 4 Measured chromosome

Quantum chromosome

$\frac{1}{\sqrt{2}}$	$\frac{1}{\sqrt{2}}$	$\frac{1}{\sqrt{2}}$	$\frac{1}{\sqrt{2}}$	$\frac{1}{\sqrt{2}}$	$\frac{1}{\sqrt{2}}$	$\cdots$	$\frac{1}{\sqrt{2}}$
$\frac{1}{\sqrt{2}}$	$\frac{1}{\sqrt{2}}$	$\frac{1}{\sqrt{2}}$	$\frac{1}{\sqrt{2}}$	$\frac{1}{\sqrt{2}}$	$\frac{1}{\sqrt{2}}$	$\cdots$	$\frac{1}{\sqrt{2}}$

⇓ *Measure*

0	1	1	0	0	1	$\cdots$	1

Binary chromosome

5.2   Initialization

In Sect. 5, we saw that there are 2^{512} possible transition functions. This huge set corresponds to the search space of the problem (all possible CA rules). Each individual of the search space is coded on a chromosome which is a binary chain of 512 bits. Initially, all qubits are set to $\frac{1}{\sqrt{2}}$ (Table 3).

5.3   Measure of Quantum Chromosomes

A measure function (Sect. 4.3) is applied on quantum chromosomes in order to extract the binary classic chromosomes, which represent the output value of the CA rule (Table 4).

5.4   Evaluation of Solutions

The next step is the evaluation of these solutions. Each binary chromosome is performed on the original image. After a few steps, the resulting image is produced. This is compared to the ground truth image, in order to determine its fitness computed with a measure of error. Three measures of error are used in this algorithm: the root mean square error (RMSE), the Hamming distance (HD), and the Structural Similarity Index (SSIM).

The Hamming distance is the number of different pixels between two images. The fitness F is computed as:

$$F = \frac{1}{1 + \text{HD}} \tag{5}$$

The RMSE is calculated according to Eq. (6)

$$\text{RMSE} = \sqrt{\frac{1}{MN} \Sigma_{r=0}^{M-1} \Sigma_{c=0}^{N-1} [E(r, c) - O(r, c)]^2} \tag{6}$$

where $O(r, c)$ is the original image (in our case the ground-truth image) and $E(r, c)$ is the reconstructed image (in our case the QGA result).

The SSIM between two images x and y is defined as [21]:

$$\text{SSIM}(x, y) = \frac{(2\mu_x\mu_y + C_1)(2\sigma_{xy} + C_2)}{(\mu_x^2 + \mu_y^2 + C_1)(\sigma_x^2 + \sigma_y^2 + C_2)} \tag{7}$$

where μ_x, μ_y are the mean of x, the mean of y. σ_x^2, σ_y^2 are respectively the variance of x, the variance of y. σ_{xy} is the covariance of x and y. Following Wang et al. [22], C_1 is set to $(0.01 * 255)^2$ and $C_2 = (0.03 * 255)^2$.

5.5 Updating Chromosomes by Interference

In the next step, the chromosomes are updated by the use of quantum interference. We update a qubit chromosome using a rotation gate $U(\theta)$. This is defined as follows:

$$U(\theta) = \begin{pmatrix} \cos(\theta) & -\sin(\theta) \\ \sin(\theta) & \cos(\theta) \end{pmatrix} \tag{8}$$

where θ is the rotation angle. This rotation has the effect of making the qubit chromosome converge to the most suitable solution. In the next step, the best solution among all the population is chosen. The i-th qubit value (α_i, β_i) is updated as:

$$\begin{pmatrix} \alpha_i' \\ \beta_i' \end{pmatrix} = \begin{pmatrix} \cos(\theta_i) & -\sin(\theta_i) \\ \sin(\theta_i) & \cos(\theta_i) \end{pmatrix} \begin{pmatrix} \alpha_i \\ \beta_i \end{pmatrix} \tag{9}$$

The parameters used in this problem are shown in Table 5. x_i and b_i are the i-th bits of x and b (the best solution). The value of $\delta\theta_i$ influences directly the speed of convergence; if it is too high, the solutions will diverge or have a premature convergence to a local optimum. The sign $s(\alpha_i, \beta_i)$ defines the direction of convergence to a global optimum. Table 5 shows the strategy of rotation of quantum gates for convergence.

5.6 Updating of Best Solutions

In this step, the better local solution and the better global solution are updated. When the algorithm is finished, the better edge obtained and the best fitness are displayed.

Table 5 Lookup table for rotation of quantum gates

$x_i\ b_i$	$f(x) >= f(b)$	$\delta\theta_i$	$s(\alpha_i, \beta_i)$			
			$\alpha_i \beta_i > 0$	$\alpha_i \beta_i < 0$	$\alpha_i = 0$	$\beta_i = 0$
0 0	False	0.005Π	−	+	±	±
0 0	True	0.005Π	−	+	±	±
0 1	False	0.08Π	−	+	±	±
0 1	True	0.005Π	−	+	±	±
1 0	False	0.08Π	+	−	±	±
1 0	True	0.005Π	+	−	±	±
1 1	False	0.005Π	+	−	±	±
1 1	True	0.005Π	+	−	±	±

6 Experimental Results

Experiments were carried out on images from the Berkeley Benchmark Dataset (BBD), which provides for each image its handmade ground truth. The population size was equal to 50. The best solution is collected within 100 generations. Figures 5 and 6 show the results of the proposed algorithm on four images from BBD: Bird, Woman, Mountain, and Landscape. It is clearly visible that the QGA method produced good edges, in comparison with the ground truth image, and a well known classical method of edge detection called Canny. Table 6 shows the mean best fitness values obtained for the three images illustrated above. For each image, the QGA and Canny edge detector are tested with the three fitness functions: Hamming distance,

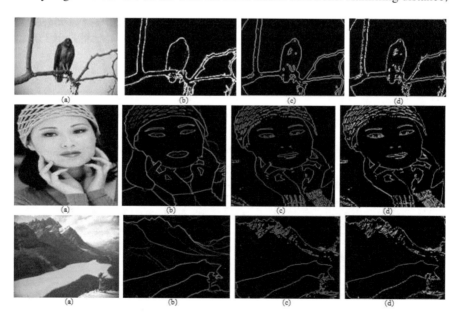

Fig. 5 Visual results of QGA and comparison. (**a**) Original image. (**b**) Ground truth. (**c**) Canny edge. (**d**) QGA edge

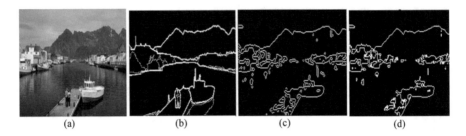

| (a) | (b) | (c) | (d) |

Fig. 6 Visual results of QGA and comparison. (**a**) Original image. (**b**) Ground truth. (**c**) Canny edge. (**d**) QGA edge

Table 6 Best fitness results for three images

Image	Iterations	QGA			Canny		
		F	SSIM	RMSE	F	SSIM	RMSE
Bird	20	$2.07e^{-4}$	0.9977	0.2767	$4.56e^{-4}$	0.9765	0.3042
	40	$2.12e^{-4}$	0.9971	0.2742	–	–	–
	100	$1.98e^{-4}$	0.9966	0.2735	–	–	–
Woman	20	$1.94e^{-4}$	0.9985	0.2482	$3.87e^{-4}$	0.9716	0.2751
	40	$1.77e^{-4}$	0.9984	0.2484	–	–	–
	100	$1.68e^{-4}$	0.9976	0.2479	–	–	–
Mountain	20	$1.59e^{-4}$	0.9992	0.2005	$3.22e^{-4}$	0.9632	0.2483
	40	$1.62e^{-4}$	0.9987	0.2132	–	–	–
	100	$1.45e^{-4}$	0.9979	0.2014	–	–	–

RMSE, and SSIM. Concerning Canny, only one iteration is sufficient to collect the fitness values. The QGA algorithm is tested over 25 runs, with respectively 20, 40, and 100 iterations.

7 Comparison Between Quantum GA and Conventional GA

In this section, a visual and numerical comparison between QGA results and CGA [10] results are presented. CGA [10] is an evolutionary algorithm which extracts a pack of rules for edge detection. It is based on a genetic algorithm which evolves CA over many generations to realize the best edge detection.

7.1 Visual Results

In the following, we consider the same images used in [10]. A QGA algorithm is applied on this set of images. The original image, the reference image, the CGA

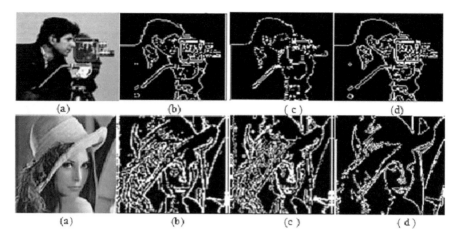

Fig. 7 Edge detection for cameraman and Lena images. (**a**) Original image. (**b**) Reference image. (**c**) CGA edge. (**d**) QGA edge

result, and QGA result are illustrated in order to show clearly the performance of QGA (Fig. 7). The result of QGA is sharply better: it allowed the extraction of all the edges in the original image with high accuracy. The visual results of the cameraman shows that the QGA algorithm has a better effect than CGA, rather more in fact as the continuity of edges is strong. QGA has a better curve outline of the edges and has good detection effects on the whole camera as well as the body features compared to CGA. This latter gives poor and discontinuous results and also includes false edges, whereas QGA gives good, clean, and almost continuous true edges. On the Lena image, the visual results clearly demonstrate that the QGA method has a better effect than CGA. CGA gives weak and discontinuous edges. It also includes false edges, whereas QGA gives clean and almost continuous and true edges.

7.2   Numerical Results

Experiments were carried out on several images for both CGA and QGA algorithms. In the conventional genetic algorithm, the population size is 100. The value of the crossover probability is 0.65. The value of the mutation probability is 0.05. The population size of QGA is 10. We recorded the best solution values after 2000 iterations, over 25 runs.

The experiment shows that QGA yielded superior results as compared to CGA. QGA gives good results even if we use a small-sized population (10 items). QGA can look for solutions near the optimum after a few iterations as compared to CGA.

The evolution of mean best fitness value (MBFV) over 2000 generations is shown in Fig. 8, for QGA and CGA algorithms. It is clear that QGA outdid CGA concerning the rate of convergence and the quality of the final results. At the

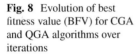
Fig. 8 Evolution of best fitness value (BFV) for CGA and QGA algorithms over iterations

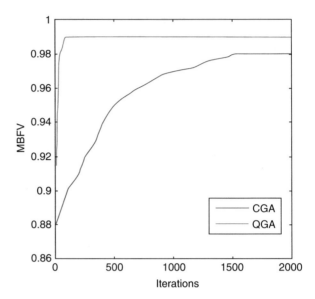

beginning of the MBFV plotting, CGA shows a slower convergence rate. After 50 generations, CGA maintains a constant convergence rate. From the beginning of the plotting, we can see that QGA has a rate of convergence faster than CGA because of its superior global search capacity. For the CGA algorithm, the best fitness value of 0.98 is achieved after about 1500 generations. Beyond this threshold, no improvement in the result is noticed. The effectiveness and applicability of QGA is demonstrated by experimental results. Figure 8 shows the high global search capacity and fast convergence of QGA against CGA.

8   Conclusion

This chapter has presented a new approach to resolving edge detection. A powerful complex system, CA, has been used to model the image. A QGA was used to explore the large search space of CA transition rules and extract efficiently the best rules that perform good edge detection. Experimental results proved that QGA is more efficient and powerful than CGA. The originality of this work mainly relates to solving edge detection by combining a quantum approach, a genetic algorithm, and CA. The main motivation behind this new method is to benefit from the parallelism of QGA in exploring the search space in order to find the best solution, with maximum effectiveness. Experimental results show that QGA is a very promising tool for exploring large search spaces like CA rule extraction, while preserving the balance between efficiency and performance. The QGA has proved its effectiveness and applicability in image processing tasks, especially for edge detection. Fast convergence and good global search capability characterize its performance. A few chromosomes are sufficient to study the problem. The use of quantum interference offers a powerful tool to reinforce the search stability. The large search space is explored, keeping the balance of efficiency and performance.

In future work, major interest will be given to comparing different QGA strategies for investigating the effect of changing rotation gate angles, the number of chromosomes, introducing crossover and mutation operators, and their impact on the performances of the QGA algorithm.

References

1. N. Abd-Alsabour, Hybrid metaheuristics for classification problems, in *Pattern Recognition-Analysis and Applications* (2016). ISBN 978-953-51-2804-5. Print ISBN 978-953-51-2803-8. https://doi.org/10.5772/65253
2. M. Batouche, S. Meshoul, A. Al Hussaini, Image processing using quantum computing and reverse emergence. Int. J. Nano Biomater. **2**, 136–142 (2009)
3. E. Casper, C. Hung, Quantum modeled clustering algorithms for image segmentation. Prog. Intell. Comput. Appl. **2**(1), 1–21 (2013)
4. L. Grover, A fast quantum mechanical algorithm for database search, in *Proceedings of 28th Annual ACM Symposium on the Theory of Computing* (1996), pp. 212–221
5. K. Han, Genetic quantum algorithm and its application to combinatorial optimization problem, in *Proceedings of IEEE Congress on Evolutionary Computation* (2000), pp. 1354–1360
6. K.-H. Han, J.-H. Kim, Quantum-inspired evolutionary algorithm for a class of combinatorial optimization. IEEE Trans. Evol. Comput. **6**(6), 580–593 (2002)
7. K.-H. Han, J.-H. Kim, On setting the parameters of quantum-inspired evolutionary algorithm for practical applications, in *Proceedings of the 2003 Congress on Evolutionary Computation* (2003), pp. 178–194
8. K.-H. Han, J.-H. Kim, Quantum-inspired evolutionary algorithms with a new termination criterion, He gate and two-phase scheme. IEEE Trans. Evol. Comput. **8**(2), 156–169 (2004)
9. T. Hey, Quantum computing: an introduction. Comput. Control Eng. J. **10**(3), 105–112 (1999)
10. O. Kazar, S. Slatnia, Evolutionary cellular automata for image segmentation and noise filtering using genetic algorithms. J. Appl. Comput. Sci. Math. **10**(5), 33–40 (2011)
11. J. Kempe, S. Laplante, F. Magniez, Comment calculer quantique? La Recherche **398**, 30–37 (2006)
12. A. Layeb, A quantum inspired particle swarm algorithm for solving the maximum satisfiability problem. IJCOPI **1**(1), 13–23 (2010)
13. A. Layeb, S. Meshoul, M. Batouche, Multiple sequence alignment by quantum genetic algorithm, in *Proceedings of the 20th International Conference on Parallel and Distributed Processing* (2006), pp. 311–318
14. A. Narayanan, Quantum computing for engineers, in *Proceedings of the 1999 Congress on Evolutionary Computation* (1999), pp. 2231–2238
15. A. Narayanan, M. Moore, Quantum-inspired genetic algorithms, in *Proceedings of IEEE Transactions on Evolutionary Computation* (1996), pp. 61–66
16. A. Rosin, Training cellular automata for image processing. IEEE Trans. Image Process. **15**(7), 2076–2087 (2006)
17. P. Shor, Algorithms for quantum computation: discrete logarithms and factoring, in *Proceedings of the 35th Annual Symposium on the Foundation of Computer Sciences* (1994), pp. 20–22
18. E.G. Talbi, Hybrid metaheuristics for multi-objective. Optim. J. Algorithms Comput. Technol. **9**(1), 41–63 (2015)

19. H. Talbi, M. Batouche, A. Draa, A quantum inspired evolutionary algorithm for multiobjective image segmentation. Int. J. Comput. Inf. Syst. Control Eng. **1**(7), 1951–1956 (2007)
20. T. Urli, Hybrid meta-heuristics for combinatorial optimization. PhD thesis, Udine University, 2014
21. Z. Wang, E.P. Simoncelli, A.C. Bovic, Multi-scale structural similarity for image quality assessment, in *Proceedings of 37th IEEE Asilomar Conference on Signals, Systems and Computers*, Pacific Grove, Nov 09–12 (2002)
22. Z. Wang, A.C. Bovik, H.R. Sheikh, E.P. Simoncelli, Image quality assessment: from error visibility to structural similarity. IEEE Trans. Image Process. **13**(4), 600–612 (2004)
23. H. Wang, J. Liu, J. Zhi, C. Fu, The improvement of quantum genetic algorithm and its application on function optimization. Math. Probl. Eng. **2013**, Article ID 730749 (2013)
24. J. Zhang, J. Zhou, H. Kun, M. Gong, An improved quantum genetic algorithm for image segmentation. J. Comput. Inf. Syst. **11**, 3979–3985 (2011)

Genetic Algorithm Implementation to Optimize the Hybridization of Feature Extraction and Metaheuristic Classifiers

Geetika Singh and Indu Chhabra

Abstract Hybridization represents a promising approach for solving any recognition problem. This chapter presents two face recognition frameworks involving the hybridization of both the feature extraction and classification stages. Feature extraction is performed through the two proposed hybrid techniques, one based on the orthogonal combination of local binary patterns and histogram of oriented gradients, and the other based on gabor filters and Zernike moments. A hybrid metaheuristic classifier is also investigated for classification based on the integration of genetic algorithms (GA) and support vector machines (SVM), where GA has been used for the optimization of the SVM parameters. This is crucial since the optimal selection of SVM parameters ultimately governs its recognition accuracy. Experimental results and comparisons prove the suitability of the proposed frameworks as compared to the other baseline and previous works.

Keywords Face recognition · Hybrid feature extraction · Support vector machine · Genetic algorithm · GA-SVM classification

1 Introduction

Face recognition technology is gaining importance primarily due to its non-intrusive nature for secured biometric identification. It has not only been found suitable for authentication and access control but has several other practical applications, such as facilitating crime investigations, carrying out secure e-commerce transactions, surveillance, finding missing people, and human–computer interaction. Though the latest professional face recognition systems have achieved a certain level of

G. Singh (✉)
Department of Computer Science and Applications, MCM DAV College for Women, Chandigarh, India

I. Chhabra
Department of Computer Science and Applications, Panjab University, Chandigarh, India

© Springer International Publishing AG, part of Springer Nature 2018
S. Bhattacharyya (ed.), *Hybrid Metaheuristics for Image Analysis*,
https://doi.org/10.1007/978-3-319-77625-5_3

accuracy, like other biometric systems their performance is limited in real-world scenarios. The quality of the fingerprint systems may be affected by the variability in the orientation of the finger or the pressure applied against the sensor. Signature verification systems may result in low accuracy due to a lack of consistency in the signatures. Accuracy of voice recognition systems may degrade by the change in voice of a person due to several factors. Similarly, robustness of facial identification systems is also challenged by various situations as the highly dynamic face objects may also undergo wide variations due to pose, lighting conditions, expression changes, occlusion, and age factors. Hence, there is a need to develop invariant algorithms that can handle these types of variations to some reasonable extent and which could provide good recognition accuracy for real-world scenarios as well. Apart from these parameters, accurate representation of faces under such situations, computational simplicity, ease of implementation, and speed are additional factors that need to be examined for an optimal recognition framework. A face recognition system, thus, needs to be robust, fast, computationally feasible, and have the capability of achieving at least equivalent or better than human recognition performance.

There are three stages in solving any face recognition problem, namely face segmentation, feature extraction, and classification. Feature extraction is considered as the most crucial stage in order to represent the face in a way that should minimize its variations and contribute to the best recognition result. Various techniques have been proposed in the literature for facial feature extraction including global, local, and hybrid methods.

Global techniques are appearance-based approaches which are applied to the whole face to extract the complete face information [4, 11, 17, 19, 30, 31, 35, 39]. These include subspace-based methods, spatial frequency techniques, and moments based methods. Subspace-based methods transform the facial image to a low-dimensional space that is eventually used for performing the recognition. These comprise some of the successful face recognition approaches like Principal Component Analysis (PCA), Linear Discriminant Analysis (LDA), Independent Component Analysis (ICA), Two-dimensional PCA (2DPCA), 2DLDA, Kernel PCA (KPCA), KLDA, and KICA [4, 11, 19, 39]. Spatial-frequency techniques include approaches such as the Fourier Transform and the Discrete Cosine Transform (DCT) [17, 35]. These map the spatial image to the frequency domain; recognition is then performed using coefficients of the low frequency band. Moments-based methods are the most widely used global face descriptors. The magnitude of these moments is used as image descriptor, as it is invariant to rotation and can be made invariant to translation and scale through proper normalization. Amongst the global methods, facial shape representation with Zernike Moment (ZM) exhibits efficient recognition ability [31, 33]. This is due to the fact that this technique is rotation invariant and therefore capable of recognizing tilted as well as posed faces. It is also invariant to noise, position, and tilt and can be made translation and scale invariant through normalization.

However, the performance of global approaches degrades in the case of large pose changes as much of the face gets occluded. Studies therefore have also emphasized local approaches, as they extract local features of the face to provide finer detail and are also effective when faces are captured at large pose angles. Local methods utilize two approaches in this context. In the first one, the image is divided into subparts and features are derived for each of these subcomponents [1, 2, 6, 12, 16, 18, 20, 38, 45]. Thus, local features are computed pixel by pixel over the input image. The second approach is concerned with the localization of different geometrical face constituents like eyes, mouth, and nose [9, 23, 42, 44]. A feature vector is obtained from the set of corresponding estimated structural parameters like ratios and distances. Much of the recent research work is centered on the former approach, as it has been found to be more robust especially in the case of variations of illumination, occlusion, and expression. A prominent descriptor in this category is Local Binary Pattern (LBP). LBP computes a histogram by considering the value of each image pixel and its neighborhood [1]. This approach has been found to yield a recognition accuracy of 97% and 79% for expression and illumination variations respectively on the FERET dataset using chi-square distance as the similarity measure. This approach has also been extended to use the AdaBoost algorithm to select discriminative LBP features which has resulted in a better recognition accuracy of 97.9% for the FERET dataset [20]. Apart from being computationally efficient and illumination invariant, LBP possesses excellent classification performance. However, it is not invariant to rotation. Orthogonal Combination of Local Binary Patterns (OC-LBP) is a recently introduced variation of LBPs with much reduced dimensionality along with better discriminative power and invariance properties as compared to the traditional LBP operator [55]. OC-LBP has achieved improvement in accuracy by up to 5% on standard texture classification datasets. Gabor filters have also proved to be effective in the case of varied facial changes [5, 36, 48]. In Gabor-based methods, features are obtained by convolving the input face image through a set of filters at different scale and orientation levels. These can efficiently represent the facial contours and edges as they are invariant to scale and orientation. However, high feature dimension and computational complexity make this technique less suitable for real-time applications. The Histogram of Oriented Gradient (HOG) is a recent descriptor that has been successfully applied to the problems of computer vision such as human detection and hand gesture recognition. Recently, it has been tested for face recognition in three variations and has provided improved results [10]. In the first case [2], a set of facial landmarks are localized using an Elastic Bunch Graph Matching (EBGM) algorithm; the corresponding HOG descriptors are computed for each of the key points. This has performed better in comparison to the classical Gabor-EBGM approach. In the second variation [12], HOG descriptors are obtained from a regular grid applied at different scales and then combined through the product rule. In the third case [38], a HOG window is placed over the entire face image to compute the feature vector. This has not only reduced the complexity but has also exhibited improved performance. Research further reveals that HOG features are robust to changes of illumination, rotation, and small displacements. In addition to this, this descriptor has lower computational complexity as compared to its competitors, which makes it suitable for real-time applications. However, both OC-LBPs and HOGs still offer considerable research space for their in-depth exploration. In the case of Gabor filters, reducing their dimensionality using suitable methods that are complementary to their behavior is again a research problem that needs addressing.

Hybrid techniques, an imitation of the human perception system, exploit the characteristics of two or more local or global approaches; these are now an active area of research. They have shown better results as the integrated feature set is capable of retaining the information about the maximum varying aspects of any facial image. Significant attempts have been made in this direction to find an optimal set of combination that can be generalized for the majority of real-world situations [14, 22, 25, 32, 33, 37, 38, 43, 46, 47, 53, 54]. Recently, a hybrid method was proposed that fuses ZM and LBP features where classification was performed at the matching score level [33]. Significantly higher accuracy was reported as compared to their individual implementation in the case of pose, illumination, and expression variations. Another promising implementation has been proposed in [38] where HOG features are extracted both at the global and local level and fused together for classification using the weighted angle distance. Although recent methods proposed in the literature focus on hybridization of feature sets, selection of an appropriate combination still remains a challenge.

For the final phase of recognizing faces, classifiers are trained to match the extracted features with their corresponding face classes as the correct outputs. Different classifiers are available, such as the distance-based, multi-layer, Back-propagation Neural Network, the Adaptive Neuro Fuzzy Inference System, and the Support Vector Machine (SVM). Selection of an appropriate classifier that is complementary with the extracted feature vector is crucial to obtain optimal recognition performance. The SVM, proposed by Vapnik, is now a well-known name among the classification strategies that is based on supervised machine learning methodology. The prime advantages of SVM include their computational efficiency, flexibility, and capability to handle large amounts of high-dimensional non-linear data. However, in order to apply this technique to specific recognition problems, its parameters, namely the regularization or the cost (c) parameter and gamma (γ), need to be adjusted using some heuristic strategy. Selection of these kernel parameters is one of the main factors affecting the application results and is a critical research area in the study of SVMs [8]. The most popular and a traditional strategy to optimize SVM parameters is based on gradient descent or grid search over the set of parameters, though it involves high computation cost, and therefore this technique is reliable only in low-dimensional datasets. The Genetic Algorithm (GA) is a global metaheuristic strategy that was first developed by John Holland in 1975. The literature suggests that this algorithm is a promising methodology over other optimization algorithms for automatically tuning the parameters of SVM as it is flexible and allows hybridization with other methodologies so as to obtain enhanced solutions [8]. In face recognition, GA has been used in conjunction with SVMs primarily for feature selection [3]. It selects a subset of extracted facial features which are fed to the SVM for final classification. However, its application to optimize SVM parameters for the problem of face recognition still needs to be investigated.

Motivated by this, the present work makes contributions both in the feature extraction stage and in the classification stage. In the feature extraction stage, two hybrid algorithms are proposed based on the fusion of distinct and complementary feature sets. In the classification stage, an improved methodology based on a hybridization of the metaheuristic GA and SVM is developed.

The first feature extraction framework is based on representing the faces with both texture as well as edge information. This is done to make the recognition algorithm memorize the combination of varying facial expressions of the individuals along with the statistically already memorized facial shapes. The texture features capture the appearance of the face such as spots and flat areas while the edges detect the face shape. Both features are unique for every individual and hence provide a mechanism for distinctive identification. Thus, the study attempts to inspect a hybrid method based on a combination of the recently introduced OC-LBP, which captures the facial texture, and HOG descriptors, which represent the facial shape.

The second hybrid implementation is achieved through the extraction of shape features for different alignments of the same face. This has helped in representing the same face in varying conditions of pose, scale, and orientation. This is achieved by extracting ZM coefficients from the Gabor filtered face images. Gabor filters represent the face image at different alignments and ZM has the capability of extracting invariant shape features. As Gabor filters are not inherently orthogonal and result in high dimensionality features, so this approach performs orthogonalization of the generated filter bank and reduces the dimensionality of the final feature vector while enhancing the discrimination power.

The classification phase is proposed to be achieved by using different distance-based metrics, including χ^2, square-chord, and extended-canberra [26], as well as SVMs with a Radial Basis Function (RBF) and χ^2 kernels. χ^2, square-chord, and extended-canberra are histogram-based metrics and may prove beneficial for the OC-LBP and HOG methods as they are based on representing the extracted feature vectors as histograms. The present study also proposes the hybridization of metaheuristic GA with a soft-computing technique, that is SVM, and explores a hybrid metaheuristic SVM model for face recognition. In this approach, the GA is used to determine the optimal values of the SVM parameters c and γ. This results in providing high classification accuracy along with better generalization ability, lower computational cost, and a fast learning speed.

The implemented techniques are evaluated for their robustness to different facial variations on benchmark face databases. A face database has also been created and the performance of implemented techniques is also verified against this database. The improved techniques have also been thoroughly compared with the state-of-the-art face recognition algorithms for establishing their recognition performance.

2 Feature Extraction

This section presents the implicit details of the underlying feature extraction techniques in the developed hybrid approaches. These include Gabor filters, LBPs, orthogonal combination of LBPs, and HOGs.

2.1 Gabor Filters

Gabor filters, as the name suggests, are capable of filtering the multi-orientational information from an image for different scales. Hence, their face description characteristics are similar to those of the human visual system. A two-dimensional Gabor filter in the spatial domain is implemented as:

$$\varphi_{u,v}(x, y) = \frac{f_u^2}{\pi \gamma \eta} e^{-\left(\frac{f_u^2}{\gamma^2} x'^2 + \frac{f_u^2}{\eta^2} y'^2\right)} e^{j2\pi x_u f'} \tag{1}$$

$x' = x \cos\theta_v + y \sin\theta_v$, $y' = -x \sin\theta_v + y \cos\theta_v$, $\theta_v = \frac{v\pi}{8}$ and $f_u = \frac{f_{max}}{2^{u/2}}$, f_{max} is the maximum frequency of the filters which is assigned to $f_{max} = 0.25$.

This definition implies that Gabor filters are complex signals generated by the Gaussian kernel functions which are further modulated by a complex plane wave whose center frequency and orientation are f_u and θ_v respectively. The values of γ and η are set to $\gamma = \eta = \sqrt{2}$. These determine the ratio between the center frequency and the size of the Gaussian envelope and, when set to some optimal value, ensure that Gabor filters of different scales for a given orientation behave as the scaled versions of each other.

To derive the Gabor face representation from a given face image $I(x, y)$, a filter bank consisting of 'u' scales with 'v' orientations is created. The image is filtered with each computed Gabor filter as:

$$G_{u,v}(x, y) = I(x, y) \times \varphi_{u,v}(x, y) \tag{2}$$

where $G_{u,v}(x, y)$ denotes the complex convolution results with both real and imaginary parts as:

$$E_{u,v}(x, y) = \text{Re}[G_{u,v}(x, y)]; \ O_{u,v}(x, y) = \text{Im}[G_{u,v}(x, y)] \tag{3}$$

respectively. Then, the phase $\varphi_{u,v}(x, y)$ and the magnitude $A_{u,v}(x, y)$ of the filter responses are computed as:

$$A_{u,v}(x, y) = \sqrt{E_{u,v}(x, y)^2 + O_{u,v}(x, y)^2}; \ \varphi_{u,v} = \arctan\left(\frac{O_{u,v}(x, y)}{E_{u,v}(x, y)}\right) \tag{4}$$

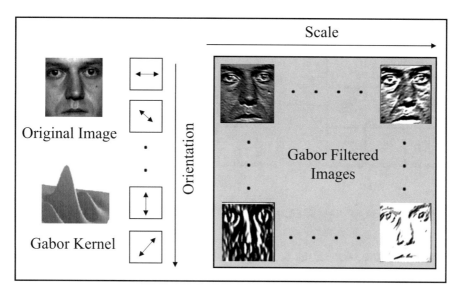

Fig. 1 Gabor filter response computation for a face image

respectively. The computed phase responses vary significantly even for the spatial locations having few pixel differences, therefore those phase features are considered unstable and hence are discarded. The magnitude responses, on the other hand, vary very slowly with the spatial position and hence are used in the final feature vector. Figure 1 demonstrates the computation of Gabor filtered images.

As each filtered response is of the same dimensionality as that of the input image, the feature vector computed is 40 times the size of the original image. To overcome this problem, the magnitude features are down-sampled using a simple rectangular sampling grid and are further projected to a low-dimensional subspace using PCA.

2.2 Local Binary Patterns and Orthogonal Combination of Local Binary Patterns

LBP is the most frequently used local descriptor which captures texture information. It is computationally simple, invariant to illumination, and provides robustness to wider pose changes. The LBP operator considers a specific neighborhood around each pixel to threshold the limits of these neighboring pixels with respect to the central pixel. The result for a neighboring pixel is 1 if its value is greater than that of the central pixel, otherwise it is 0 as inferred by Eq. (5) and demonstrated in Fig. 2.

$$b\left(p_i - p_c\right) = \begin{cases} 1, & p_i \geq p_c \\ 0, & p_i < p_c \end{cases} \tag{5}$$

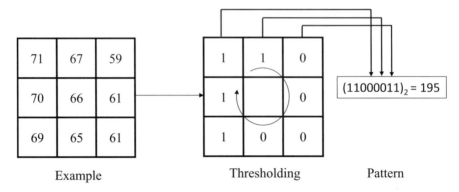

Fig. 2 Pixel-wise generation of LBP codes

where p_c is the value of the central pixel and p_i represents gray values of the p neighboring pixels. The LBP code for the central pixel is then computed as:

$$\text{LBP} = \sum_{i=0}^{p} b\,(p_i - p_c)\,2^p \qquad (6)$$

The final LBP description of an image is derived by computing the LBP code of each pixel to build a histogram.

The chief benefit of LBPs is that they are invariant to illumination variations and are very fast to compute. The main drawback, however, is the high dimensional histograms produced by the LBP codes. For instance, the LBP description for 3×3 neighborhoods results in a 2^8, that is a 256-dimensional histogram. Uniform-LBP [45] is an alternative to this original LBP operator which significantly reduces the number of LBP histogram bins from 256 to 59 in the case of eight-bit patterns. This is due to the fact that out of 256 patterns, only 58 uniform patterns preserve 90% of texture information while the remaining patterns mostly depict noise. Center-symmetric LBP [18], called CS-LBP, compares only center-symmetric pairs of pixels instead of comparing each pixel with the central pixel. Hence, this halves the number of comparisons as compared to the original LBP operator. OC-LBP [55] has been proposed recently to reduce dimensionality by considering fewer neighboring pixels while retaining the discriminative and photometric invariance properties. It takes into account two different non-overlapping four-orthogonal-neighbor operators and combines their histograms. The first operator is obtained by considering only horizontal and vertical neighbors as in Eq. (7) and the second operator is obtained by considering the diagonal neighbors as in Eq. (8). The final descriptor is then derived by combining the histograms of the first and second operators (Eq. (9)). The size of the LBP is reduced significantly to $2^4 \times 2 = 32$

in comparison to 256 by the original operator.

$$\text{OC-LBP1} = 2^0 \times b\,(p_0 - p_c) + 2^1 \times b\,(p_2 - p_c)$$
$$+ 2^2 \times b\,(p_4 - p_c) + 2^3 \times b(p_6 - p_c) \qquad (7)$$

$$\text{OC-LBP2} = 2^0 \times b\,(p_1 - p_c) + 2^1 \times b\,(p_3 - p_c)$$
$$+ 2^2 \times b\,(p_5 - p_c) + 2^3 \times b(p_7 - p_c) \qquad (8)$$

$$\text{OC-LBP} = [\text{OC-LBP1}, \text{OC-LBP2}] \qquad (9)$$

2.3 Histogram of Oriented Gradients

These shape descriptors express the local object appearance and shape through the distribution of local intensity gradients and edge orientations. At first, both the x- and y-directional gradients G_h and G_v of the input image are computed using the Sobel filter. Then, the magnitude $M_G(x, y)$ and orientation $\theta_G(x, y)$ of these gradients are obtained where each pixel is represented as a gradient vector consisting of both magnitude and direction:

$$M_G = \sqrt{G_h(x, y)^2 + G_v(x, y)^2} \qquad (10)$$

and

$$\theta_G = \arctan\left(\frac{G_h(x, y)}{G_v(x, y)}\right) \qquad (11)$$

Afterwards, the image is divided into small connected areas called cells and a histogram of edge orientations is obtained for each of these spatial cells. To achieve this, the gradient angles of all pixels in each of the cells are quantized into a number of bins B so that the magnitudes of identical orientations can be accumulated in the form of a histogram as depicted in Fig. 3.

Orientation bins are evenly spaced over 0–180° for unsigned gradient evaluation and vary from 0° to 360° for signed gradient evaluation. The length of the histogram vector of each cell is indicated by the number of bins used. To make the final descriptor invariant to illumination and contrast, the histogram of each cell is normalized. This is performed by estimating the measure of intensity over larger spatially connected blocks and utilizing these results to normalize each cell within that block. The final HOG feature vector is then obtained by concatenating normalized cell histograms for all these blocks.

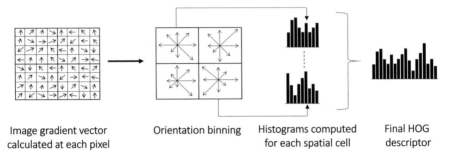

Image gradient vector   Orientation binning   Histograms computed   Final HOG
calculated at each pixel   for each spatial cell   descriptor

Fig. 3 HOG feature computation and extraction

3 Distance-Based Classification

For the final phase of recognizing faces, classifiers are trained to match the extracted features with their corresponding face classes as the correct outputs. In the present study, both the distance-based classifiers and the SVM have been utilized. The distance-based classifiers used are elaborated in this section. Section 4 discusses the proposed hybrid SVM model utilizing the metaheuristic GA technique.

Distance-based classifiers, called statistical classifiers, are well-known and the oldest and simplest methods used for classification. They estimate the similarity between two feature vectors of database and query images through some predefined function. The χ^2, extended canberra, and square-chord distances are histogram-based metrics which work efficiently for histogram features such as those generated by LBP and HOG. The extended canberra metric was recently proposed in the literature [26] and has shown improved results as compared to both χ^2 and square-chord metrics. The distance between two sets of n-dimensional feature sets x and y of database and query images respectively are defined as:

Chi-square distance:

$$d = \sum_{i=1}^{n} \frac{(x_i - y_i)^2}{(x_i + y_i)} \qquad (12)$$

Extended canberra distance:

$$d = \sum_{i=1}^{n} \frac{|x_i - y_i|}{|x_i + u| + |y_i + v|}; u = \sum_{i=1}^{m} x_i/n, v = \sum_{i=1}^{m} y_i/n \qquad (13)$$

Square-chord distance:

$$d = \sum_{i=1}^{n} (\sqrt{x_i} - \sqrt{y_i})^2 \qquad (14)$$

4 Proposed Hybrid Metaheuristic GA-SVM Model for Classification

SVM is an advanced AI-based machine learning approach which possesses good generalization ability. It has been found to be one of the most powerful tools for solving any kind of classification problem. However, one of the main drawbacks is that the level of accuracy achieved depends extensively on the parameters chosen to design the SVM model. Thus, it becomes necessary that the parameters of the SVM classifier are configured properly so as to utilize effectively its capability. One of the most widely used approaches for optimizing the SVM parameters is based on gradient descent or the grid searching approach. This technique is based on an exhaustive search of the defined subset space for the parameters. This subset space is specified using a lower bound and an upper bound value along with the number of steps to iterate through that space. The performance of each combination of the parameter values is established using some performance metric. The grid-search approach prevents the over-fitting of data; however, it has certain limitations. One possible drawback is that this method is time consuming as the number of possible combinations to be evaluated grows exponentially with the increase in the number of parameters to be optimized. In addition, this technique also has a tendency to converge to only a sub-optimal or a locally optimal solution. Hence, alternative techniques are required that can show precedence over the traditional method. For such a requirement, GA-based methods have been proved in some studies to be a better choice to determine the parameters [3, 8, 24, 50]. This chapter investigates GAs to tune automatically the parameter of SVM for face classification.

4.1 Support Vector Machines

SVM is a useful technique for non-linear data classification. It separates a non-linear separable classification problem by mapping the data to a linearly separable feature space using a non-linear map. This mapping is achieved by the use of a kernel function. In this study, the RBF and the χ^2 kernel functions are utilized for designing the SVM model. This is due to the fact that these two kernel functions can analyze high-dimensional non-linear data and require only two parameters to be optimized, namely c and γ. In addition, the χ^2 kernel proves effective when the features are histogram-based. The RBF and the χ^2 kernel are defined as:

$$\text{RBF} = e^{-\gamma \times |u-v|^2} \tag{15}$$

where u, v are feature vectors and $|u - v|^2$ is the squared Euclidean distance between u and v,

$$\text{chi-square} = e^{-\gamma (\chi^2)} \tag{16}$$

where χ^2 is the chi-square distance between two feature vectors.

The two important normalization parameters for these kernels are the cost (c) and the gamma (γ). c controls the trade-off between achieving a low training error and a low testing error, that is it is the ability to generalize the classifier to the unseen data. γ determines the width of the bell-shaped Gaussian surface. Larger values of gamma result in over-fitting while smaller gamma values cannot capture the 'complexity' or 'shape' of the data effectively.

4.2 Genetic Algorithm

GA, first proposed by John Holland in 1975, is a method for solving optimization problems through the process of Darwinian natural selection and genetics in biological systems. Unlike a grid search algorithm which can work ambiguously, GA can find the optimal solution for large-scale permutation problems very efficiently. GA starts with a set of candidate solutions called a population and each solution is represented as a chromosome. This obtains the optimal solution, that is the solution of the problem through a series of iterative computations. For this, the successive populations of alternate solutions are generated by the application of reproduction operators (namely, crossover and mutation) on the chromosomes. A crossover operator exchanges the genes between two chromosomes using a crossover point and, in mutation, the genes are altered, that is a changing of the binary codes of the genes. The quality or fitness of the resulting chromosomes at each step is determined through a fitness function. The fit chromosomes have a higher probability of being selected into the recombination pool and forming the population for the next step. This process is repeated until acceptable results are obtained, that is the termination conditions are satisfied.

One of the prime advantages of GA is that it offers flexibility of searching even when the range and other dependencies of the SVM parameters are not known at all. It also allows hybridization with other methodologies in order to obtain improved solutions [8]. These advantages make GAs reasonable candidates for overcoming the disadvantages of SVM.

4.3 Chromosome Design

The design of chromosomes is an important step in the application of GAs to solve any optimization problem. As only c and γ need to be optimized in our study, the chromosome consists of two parts, one corresponding to each parameter. Each chromosome is represented as a bit string using a binary coding system as shown in Fig. 4.

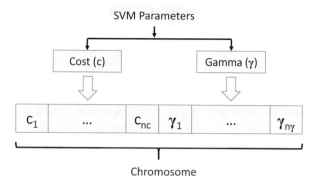

Fig. 4 Encoding of chromosome

$c_1 - c_{nc}$ represents the value for parameter c where nc is the number of bits representing c. $\gamma_1 - \gamma_{n\gamma}$ is the value for parameter γ and $n\gamma$ is the number of bits used to represent γ. These bit strings representing the genotype of the parameters c and γ are then converted to phenotypes which is an inverse process of encoding and coverts the binary chromosomes to their corresponding numeric values. This conversion is accomplished as follows:

$$p = \min_p + \frac{\max_p - \min_p}{2^n - 1} \times d \qquad (17)$$

where p is the phenotype of the binary string, $\max_p$ is the maximum value of the parameter, $\min_p$ is the minimum value of the parameter, d is the decimal value of the bit string, and n is the length of the binary string.

4.4 Fitness Function

Fitness function values of the chromosomes determine the effectiveness of the corresponding c and γ parameters in designing a more generalized and reliable SVM model. One of the widely used performance metrics to assess the generalization ability of the SVM classifier is k-fold cross-validation. On a given training dataset, the higher the cross-validation classification rate, the greater is the SVM generalization ability. Thus, in this study, the k-fold cross-validation approach has been used to estimate the fitness values. In this technique, training data is randomly divided into k subsets. The classifier is trained on the $k - 1$ subsets and tested on the remaining kth subset. The training process is repeated for k iterations and the final classification accuracy is obtained as the average of all the k classification rates.

The k-fold cross-validation fitness function F is defined as:

$$F = \frac{1}{k} \sum_{i=1}^{k} R_i \tag{18}$$

where R_i is the classification rate of the ith iteration. The literature reports that tenfold cross-validation results in optimal computation time and variance [50]. Thus, in this study, a tenfold cross-validation approach has been used to assess the performance of the designed SVM model.

4.5 Design of the Proposed GA-SVM Model

The complete methodology followed for implementation of the proposed GA-SVM parameter optimization process is shown in Fig. 5 and the steps are summarized as follows:

1. Scale the training data (i.e. the extracted feature vectors of the training images) for transforming it to the format of an SVM package. This avoids those attributes which are in greater numeric ranges from dominating those in smaller numeric ranges and also avoids numerical difficulties during the calculation. This results in higher accuracy rates. Each feature is scaled as follows:

$$f' = \frac{f - f_{min}}{f_{max}} \tag{19}$$

   where f' is the scaled feature value, f is the original value, f_{min} is the minimum value of that feature, and f_{max} is the maximum value.
2. Transform the parameters c and γ as chromosomes and represent them as binary strings. Further, initialize the population of N chromosomes.
3. Decode each chromosome, that is convert each genotype to a phenotype.
4. For each chromosome represented by c and γ, evaluate the tenfold cross-validation accuracy of the SVM classifier using the training data. This step yields the classification accuracy for a given set of parameters.
5. Select the chromosomes (i.e. c and γ values) with respect to the greater fitness values or the cross-validation accuracies.
6. If the termination condition is met, stop the process. In the present study, the termination criterion is the maximum number of generations to be evolved. If the condition is not satisfied, produce the next generations using the reproduction operators of crossover and mutation.

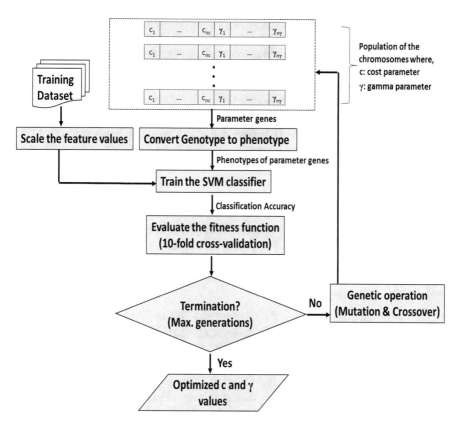

Fig. 5 Architecture of the proposed GA-SVM method

5 Proposed Hybrid Face Recognition Approaches

Two hybrid face recognition algorithms utilizing the complementary properties of the existing feature extraction approaches and the GA-SVM classifier have been designed and analyzed. The present section explains their strategic execution along with the detailed step-by-step mechanism followed for their implementation.

5.1 Integrating OC-LBP and HOG Features

LBP captures texture information and is invariant especially to monotonic light changes and is computationally very simple. However, it is not robust to geometric transformations and cannot capture the local shape and edge information. HOG on the other hand can effectively represent the edges as well as the facial contours, apart from its computational simplicity and robustness to illumination changes and geometric transformations. Though it can represent the edge information it cannot capture the texture information and its performance also degrades when the facial edges are noisy. LBP, however, has the capability to filter out those noises from the image and thus has proved to be complementary in this aspect.

Hence, the combination of HOG and LBP has been attempted in the literature for human detection [41], palm tracking [15], and object localization [49] with good results. Following this line of research, a method based on the combination of these two complementary feature sets has been tested in the present study using OC-LBP in place of LBP due to its better reported performance. OC-LBP reduces the time complexity as well as the dimensionality of the LBP operator and also increases the discriminative power of the extracted feature set. As the number of OC-LBP and HOG features generated are approximately equal, it has become feasible to integrate these two features without any domination of one over the other. Also, both the descriptors are very fast to compute. Thus, the total time for feature extraction is extremely low which makes this framework suitable especially for a real-time environment.

The procedure followed for the proposed approach is depicted in Fig. 6. It includes the following steps:

1. Normalize an image to 64×64 standard size. Divide the image in 64 blocks of size 8×8 pixels and compute HOG and OC-LBP codes for each block.
2. Compute a histogram for each of the OC-LBP and HOG blocks. Combine the histograms of all the blocks into HOG and OC-LBP feature vectors respectively.
3. Normalize the histograms using L_2-norm. This is required to map the feature sets onto a common range so that they can easily be combined and none of them dominates over the other. The number of HOG features obtained per image $= 1764 = 49$ overlapping blocks $\times$ 36 features $= 49 \times$ (9 bins $\times$ 4 cells/block) features. The number of LBP features obtained per image $= 2048 = 64$ cells $\times$ 32 features from each cell.
4. Integrate the HOG and OC-LBP feature vectors.
5. Obtain the matching scores between the integrated feature vectors using χ^2, extended canberra, and square-chord histogram-based distance metrics as well as the GA-SVM algorithm with a χ^2 kernel. Distance-based classifiers, as taken in the literature, are considered so that this method can be compared with the LBP and HOG based methods presented in the literature with the same evaluation protocol. SVM-based classification is proposed for this framework as it complements the feature sets and provides high recognition accuracy.

5.2 Gabor Filtered Zernike Moments

ZM polynomials are inherently orthogonal to one another and thus represent the global characteristics of an image without any redundancy and overlap. Gabor filters capture the local multi-orientational information at different scales from the

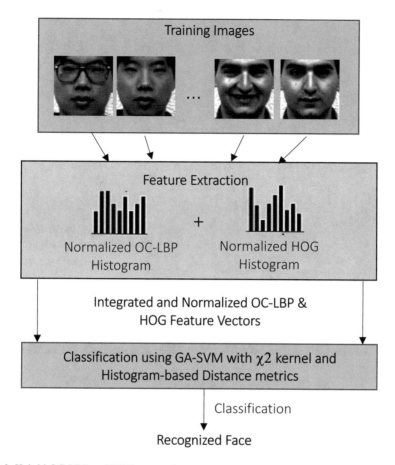

Fig. 6 Hybrid OC-LBP and HOG approach

facial image. They exhibit desirable properties of orientational selectivity and spatial locality; however, they are not orthogonal which makes the information extracted by them highly redundant. This duplicity of information in the extracted feature vector may affect the recognition accuracy of the classifier. In addition, the Gabor face representation is of very high dimensionality in that the computed feature space is about 40 times more dimensional for five scales and eight orientations than that of the original image space.

Thus, it is essential to alter the Gabor features in a way that derives a compact, discriminative, and non-redundant face representation. The study proposes to extract ZM coefficients from the Gabor filtered images. This orthogonalizes the generated filter bank and also reduces the dimensionality of the final feature vector. The orthogonalization of the filters eliminates the redundancy of information and increases their discrimination power. It also enables ZM to behave as a local image descriptor and extract the finer details at different scales and orientations. The

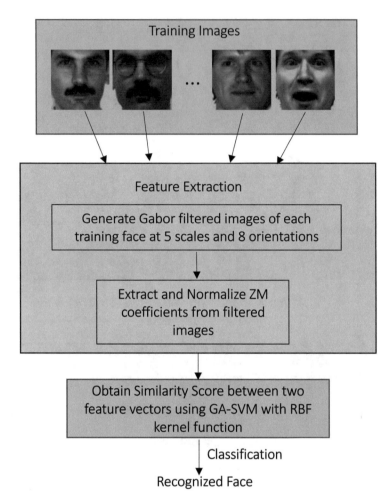

Fig. 7 Gabor filtered Zernike moments

final descriptor, thus, possesses the properties of orthogonality, compactness, and discrimination. The procedure for this proposed scheme is depicted in Fig. 7. The complete procedure comprises of the following steps:

1. Resize an image to 64 × 64 pixels standard size. Generate the Gabor filter bank of 40 Gabor filters at five scales and eight orientations and convolve the input face image with each filter of the generated filter bank. This results in 40 Gabor filtered images.
2. Compute the ZM descriptors from each of these Gabor filtered images up to an optimal order. The order for extracting ZM features is derived through experimentation.

3. Accumulate the ZM descriptors to accommodate all the filtered images into a single feature vector and normalize it using L_2-norm.
4. Measure the similarity score among feature vectors of training face images as well as the test face image using a GA-SVM classifier with an RBF kernel function to provide the output face.

6 Empirical Evaluation

The developed methodologies were implemented through MATLAB code in a Microsoft Windows environment on a Pentium PC with a 2.93 GHz CPU and 2 GB RAM. SVM classification was realized using the LibSVM package [7]. Techniques have been evaluated for their robustness to different face variations, such as facial expressions, aging, pose, and illumination.

6.1 Datasets Used

Experiments were performed on various datasets of the standard ORL, Yale, and FERET databases. These datasets include a range of variations and comprise randomly selected training and test images. The results presented in the following sections therefore are the mean of several possible experiments carried out to make the analysis at least a good representative.

Results of the developed face recognition methodologies have been compared with similar and state-of-the art approaches for their validity. The applicability of these techniques in a real-time scenario has been determined through their computational complexity analysis.

6.1.1 ORL Database

AT&T/ORL database consists of 400 face images for 40 individuals as depicted in Fig. 8. There are 10 images in Portable Gray-Map format for each person. Images are in the grayscale and of size 92×112 pixels each. This database contains pose, expression, lighting, occlusion, scale, and position variations. Poses range from $0°$ (frontal) to $\pm 20°$. Expression variations are presented in a way that either a person is smiling or not smiling and has open or closed eyes. There are also small light variations and occlusion of the eyes with eye-glasses. In this database, 10 images of each individual have been split into three groups containing three, four, and five randomly selected images per person respectively in the training set and the remaining images in the test set as considered in the literature.

Occlusion   Scale

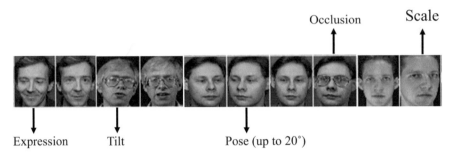

Expression   Tilt   Pose (up to 20°)

Fig. 8 Sample face images from ORL face database

Expression

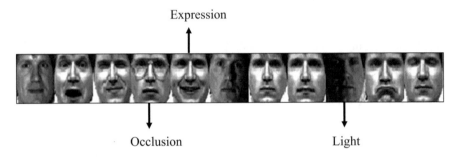

Occlusion   Light

Fig. 9 Face images of one subject from the Yale face database

6.1.2 Yale Database

The Yale database has 165 total face images for 15 persons. There are 11 images per person in grayscale form as shown in Fig. 9. Graphics Interchange Format (GIF) has been used to store the images. Each image is of the size 243×320 pixels. This database presents major variations of light and facial expressions. Therefore, it is a challenging database for checking the robustness of face recognition algorithms especially in the case of light changes. To incorporate the light changes, images have been captured by placing the light source at the left, right, front, and back of each individual. Images have also been captured with different expressions such as sad, happy, and with the mouth open. There are also occlusions of the eyes with eye-glasses. For this database, experiments have been performed on different groups consisting of three, four, five, and six randomly selected images for the training set and the remaining eight, seven, six, and five images respectively for the test set.

6.1.3 FERET Database

The FERET database is one of the most challenging datasets for the face recognition research community. It has 14,051 grayscale images of size 256×384 pixels for 1196 individuals. In the literature, this database has been widely used by well-known face recognition methods for the testing and validation of their results.

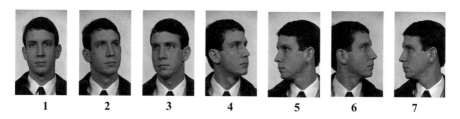

<table>
<tr><td>1</td><td>2</td><td>3</td><td>4</td><td>5</td><td>6</td><td>7</td></tr>
</table>

Fig. 10 Face images of one subject from experimental Set 1

<table>
<tr><td>ba</td><td>bb</td><td>bc</td><td>bd</td><td>be</td><td>bf</td><td>bg</td><td>bh</td><td>bi</td><td>bj</td><td>bk</td></tr>
</table>

Fig. 11 FERET 'b' category images of one subject

It has wide variations in poses ($\pm 90°$), expressions, illumination, occlusion as well as aging. On this database, experiments have been performed on a subset of randomly selected face images of 100 persons in seven different poses of $\pm 0°$, $\pm 22.5°$, $\pm 67.5°$, and $\pm 90°$ referred to as Set 1 (Fig. 10) for the study. For experimentation, three images per person are selected for training and the rest considered for testing. This results in 300 images for training and 400 images for testing.

Experiments have also been performed on the database's category 'b' images of 200 subjects, named Set 2 in the study, as shown in Fig. 11. For each subject, there are 11 images from ba to bk where ba represents the frontal pose, neutral expressions, and no light variation. Images bb through be are with pose angles of $-15°$, $-25°$, $-40°$, and $-60°$ and are symmetric analogues of images bf to bi which are with pose angles of $+15°$, $+25°$, $+40°$, and $+60°$. The bj image was captured with different expressions and bk under different light conditions.

The FERET evaluation protocol also partitions the database into a gallery set: fa with 1196 frontal face images and four test sets; fb (1195 images with facial expression variations); fc (194 images with illumination variations); dupI (722 images); and dupII (234 images). The dupI and dupII sets depict age variations. Sample face images from these datasets are shown in Fig. 12. Experiments were also performed on these FERET gallery and probe sets.

For comparative analysis, images in all these three databases are normalized to the standard size of 64×64 pixels.

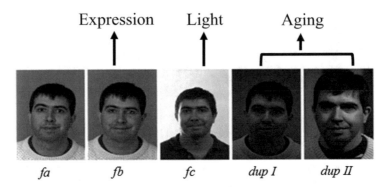

Fig. 12 Face images from the FERET gallery/probe sets

6.2 Implementation Parameters

The main parameters of the GA employed in this study are: population size (100), maximum number of generations (500 or the fitness value does not improve during the last 100 generations), probability of crossover (0.7), elite count (2), and mutation rate ($=0.02$). Tournament selection has been used as the selection strategy over the roulette wheel selection as it does not require sorting the population of chromosomes by their fitness values. Among the crossover techniques, uniform crossover has been used in which each gene of the child chromosome is selected randomly from the corresponding genes of the two parents. For the mutation, the genes are altered, that is changing the binary gene code from 0 to 1 or vice versa. The search range of parameter c is set to $[2^{-1}, 2^{12}]$, while the search range of parameter γ is set to $[2^{-12}, 2^2]$. The study evaluates the results of both the grid-search based and the proposed GA-based SVM model. The steps followed to achieve the SVM classification using grid search are briefly listed below:

1. Scale the training data to avoid domination of greater numeric ranges over smaller numeric ranges.
2. Perform a 'grid search' using a tenfold cross-validation technique to find the best values for c and γ and use these parameters to train the complete training set. During this process, various pairs of (c, γ) values are tried and the one with the best cross-validation accuracy is selected. The range for c and γ is established as $c = 2^{-5}, 2^{-3}, \ldots, 2^{15}$, and $\gamma = 2^{-15}, 2^{-13}, \ldots, 2^3$ for an exhaustive search.
3. Test the classifier on the new unseen data to verify its performance.

The recognition rate (in percentage) is measured as:

$$\text{Recognition rate} = \frac{N_t - N_r}{N_t} \times 100 \qquad (20)$$

where N_t is the total number of images in the test set and N_r is the number of incorrectly recognized images.

6.3 Database Generation for Validation

After establishing the efficacy of the developed techniques on benchmark databases, a self-designed database is created and utilized to validate their performance in-house. This database not only evaluates the developed techniques under diverse pose and light conditions but also verifies their performance for a real-life scenario. The collected database comprises 15 images per person for a total of 10 individuals. Each image has a size of 128×128 pixels and is stored in the gray-scale format as a joint photographic experts group (JPEG) file. This file format is chosen as it is efficient in storing photographic gray-scale images with less storage along with high quality.

The individuals are captured under varying light and pose conditions. The poses range from $\pm 10°$ to $\pm 60°$ ($\pm 10°$, $\pm 20°$, $\pm 30°$, $\pm 45°$ and $\pm 60°$), thus there are a total of 10 pose images for each person. These posed images are captured by placing the camera in the respective directions. Posed images from this database are shown in Fig. 13.

Light variations are captured through four images, with the light source placed at the left, right, front, and back of the person being photographed. A neutral frontal face image at $0°$ (no pose variation) with no light variation is also taken for each person. The images with light intensity variations are captured by placing the light source at the desired position. The corresponding image set is shown in Fig. 14.

The entire facial set of each individual is captured in one sitting. Care is taken to ensure uniform facial expressions of each person throughout the sessions. The illumination of the photography room is also kept constant.

6.4 Performance Evaluation of the Integrated OC-LBP and HOG Approaches

The performance of LBP, OC-LBP, and OC-LBP+HOG methods is evaluated on experimental databases for different kinds of facial variations. Tables 1, 2, and 3 present the results on the FERET, Yale, and ORL databases respectively.

Table 1 shows that the highest recognition rate of 99.5%, 99.7%, 96.2%, and 93.8% are obtained with the OC-LBP+HOG-GA-SVM approach on the fb, fc, dup I, and dup II datasets respectively. On Set I and Set II, the highest recognition rates of 95.6% and 97.9% are obtained by the proposed approach. It is further observed that on these two datasets both the HOG and OC-LBP descriptors show similar performance with HOG depicting a slight improvement of the maximum up to 1%.

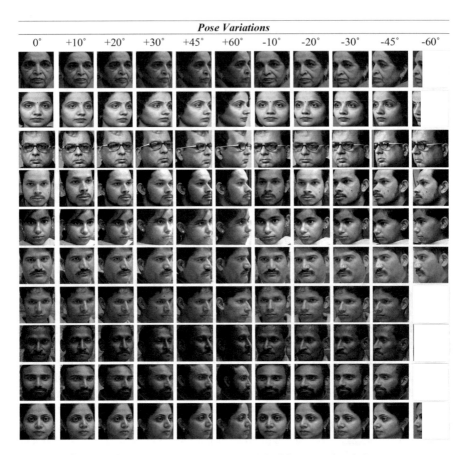

Fig. 13 Self-generated database: face images captured in different pose variations

Fig. 14 Self-generated database: face images captured in different light variations

Table 1 Performance of OCLBP+HOG method on the FERET database

Method	fb (expression variations)				fc (light variations)			
	χ^2	Square chord	Extended canberra	GA-SVM	χ^2	Square chord	Extended canberra	GA-SVM
LBP	96.3	97.5	97.4	98.7	76.8	77.2	77.1	85.4
OC-LBP	97.3	97.9	98.1	99	77.1	79.8	79.9	87.6
HOG	91.2	93.1	93.2	98.8	82.5	84.3	84.7	93.1
OC-LBP+HOG	98.7	99.1	99.2	99.5	96.4	97	97.1	99.7
	dup I (age variations)				dup II (age variations)			
LBP	68	70.2	70.3	77	61	63.2	63.3	71.9
OC-LBP	69.4	72.4	72.6	79.2	62.5	64	63.8	74.2
HOG	85.9	86.2	86.3	94.1	75.5	76.9	77.2	85.3
OC-LBP+HOG	87.1	87.7	87.9	96.2	82.3	83.6	83.8	93.8
	FERET (Set 1)				FERET (Set 2)			
LBP	72.6	73.2	72.9	77.6	80.6	82.3	82.9	88.6
OC-LBP	72.1	73.4	73.2	78.9	82.3	84.4	84.3	88.9
HOG	73.3	74.5	74.7	78.2	83.3	84.5	84.6	89.1
OC-LBP+HOG	89.0	90.3	91.5	95.6	88.4	90.1	91.2	97.9

Table 2 Performance of the OC-LBP+HOG method on the Yale database

Method	Training images per person							
	3				4			
	χ^2	Square chord	Extended canberra	GA-SVM	χ^2	Square chord	Extended canberra	GA-SVM
LBP	90.6	91.1	91.8	95	92.8	94.2	94.8	97
OC-LBP	93.2	94.7	95	98.1	94.2	96.1	96.3	97.9
HOG	93.7	95.1	95.2	97.9	95	96.9	96.8	97.5
OC-LBP+HOG	96	97.5	97.9	98.7	97.5	98.9	98.7	98.9
	Training images per person							
	5				6			
	χ^2	Square chord	Extended canberra	GA-SVM	χ^2	Square chord	Extended canberra	GA-SVM
LBP	93.1	94.5	94.9	97	93.7	95.1	95.6	97.1
OC-LBP	95.3	96.6	96.7	97.9	95.8	97	97	99
HOG	95.8	96.1	96.6	98.3	96	97.9	97.8	98.9
OC-LBP+HOG	98	98	98.1	99	98.5	99.1	99.2	99.7

This is due to the fact that higher pose angles occlude a significant portion of the face image and both features, due to their local nature, are robust to such distortions. On comparing the results of different distance metrics, it is observed that square-chord and extended-canberra metrics increase the performance accuracy by 3% and 3.2%

Table 3 Performance of the OC-LBP+HOG method on the ORL database

Method	Training images per person											
	3				4				5			
	χ^2	Square chord	Extended canberra	GA-SVM	χ^2	Square chord	Extended canberra	GA-SVM	χ^2	Square chord	Extended canberra	GA-SVM
LBP	89	91.5	92	97	92.5	94	94	97.5	95.5	97	96.9	99.5
OC-LBP	90.5	93	93	98	94.5	96	95.5	98	96.5	98	98	99
HOG	90	92.5	93	99	93	94	94.5	98.5	96	97	97.1	99
OC-LBP+HOG	95.5	97	97.5	99.5	98.5	99	99	99.5	99.5	99.5	99.5	99.5

respectively over the χ^2 metric. GA-SVM provides superior recognition rates and exhibits an improvement of up to 11% over χ^2, square-chord as well as extended-canberra distance metrics.

The dup I and dup II sets are extremely challenging datasets as they include not only the age variations but also contain lighting, expression, and background variations. Also, the performance of the LBP descriptors is better than HOG for the fb dataset (expression variations) whereas HOG performs better than LBP for fc (light changes) as well as the dup I and dup II (aging) datasets. This is because HOG features capture the edge information which may degrade their performance in the case of expression changes, though that has proved effective for light changes and aging.

It is observed from Table 2 that on the Yale database the recognition rates obtained from the OC-LBP+HOG-GASVM approach are higher than those obtained by OC-LBP and HOG by up to 5.5% and 5% respectively. It is well known that both the HOG and LBP methods are invariant to light changes which contribute to the high results on this database. It is, however, noted that the HOG method shows more robustness to image intensities than LBP. This is due to the fact that HOG computes the image gradients which capture the edge information and thus provide more resistance. Normalizing each cell histogram also introduces better invariance to light, shadows, and even edge contrast.

On comparing the results of different distance metrics on all datasets of the Yale database, both the square chord and extended canberra metrics are found to provide a performance improvement of up to 1.9% and 2.1% respectively over the χ^2 metric. GA-SVM further supplements the recognition rate up to 8%. From the results it can therefore be concluded that the integrated OCLBP+HOG-GASVM method is robust to illumination as well as for expression variations.

For the ORL database (Table 3), in the case of four and five images in the training set and for the remaining six and five images respectively in the test set, an average recognition rate of 99.5% is achieved using the SVM classifier. The performance of HOG features is better than LBP on this database due to the presence of pose, tilt, and scale variations; HOG features are more robust to such displacements.

6.5   Performance Evaluation of the Gabor Filtered ZM Method

The efficacy of the Gabor filtered ZM approach (GFZM) is determined and compared to the individual ZM and Gabor techniques. ZM features are extracted from Gabor filtered images of a face. This filter inherently possesses invariance to illumination changes and also extracts features invariant to pose and scale. In other words, it pre-processes the face images. This implies that the extraction of ZM features becomes independent of the database under consideration and becomes confined to only the uniformly filtered face images across multiple databases. Figure 15 shows the implication of this observation on the Yale database. It

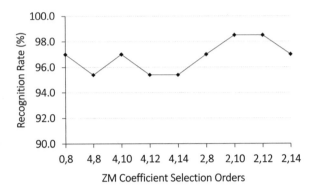

Fig. 15 Effect of GFZM order selection on the Yale database

Table 4 Performance of GFZM on the ORL and Yale databases

Number of randomly selected images in the training set	Descriptor			
	Gabor filter (established)		GFZM-GA-SVM (proposed)	
	ORL	Yale	ORL	Yale
3	95.1	94.9	97.5	96.2
4	96.7	95.4	99.5	97.3
5	97.2	97.9	99.8	99.5
6	-	98.7	-	99.5

Table 5 Performance of GFZM on the FERET database

	Dataset	Descriptor	
		Gabor filter (established)	GFZM-GA-SVM (proposed)
FERET database subsets	Set 1	83.3	88
	Set 2	82.1	93.2
FERET gallery/Probe sets	fb	75.2	88.5
	fc	83.7	99.3
	dup I	69.3	76.4
	dup II	67.4	75.8

illustrates the recognition accuracy obtained with different groups of orders of ZM. On this dataset, the traditional ZM features have to be extracted for orders (4, 14) [31]. However, the figure shows that orders (2, 10) are more beneficial for extracting ZM features from Gabor filtered face images. The same behavior is observed in all experimental databases. Thus, ZM order of (2, 10) are selected for further experimental analysis.

The performance is evaluated in Tables 4 and 5 for the ORL, Yale, and FERET databases.

The Gabor filter is actually a linear filter used for edge detection. Gabor representation of the facial image filters out the facial contours and the edges. It is observed from Table 4 that on the ORL and Yale databases this property of Gabor filters is proved effective for recognition under different illumination conditions. Further processing of the Gabor filtered images with ZM increases the robustness to light and contrast. From the results, it is observed that GFZM-GA-SVM achieves the highest recognition of 99.8% and 99.5% for the ORL and Yale databases respectively.

In the case of the FERET database (Table 5), the results of Set 1 and Set 2 datasets show that the Gabor filtered ZM approach has proved tough especially to pose variations. However, Gabor images cannot effectively extract expression invariant features and ZM cannot represent age changes, thus the results on the fa/fb, dup I, and dup II datasets are not as competent as expected.

7 Performance Comparison with Other Similar and State-of-the-Art Methods

In this section, a comparison among the combined methods and other similar state-of-the-art approaches is made for the three benchmark databases.

It is inferred from Table 6 that on the ORL database the recognition rate of the integrated approaches is higher as compared to that of the similar and state-of-the-art methods available in the literature. It is noteworthy that the GFZM-GA-SVM approach achieves the highest recognition rate of 99.8%. In the literature, the reported highest recognition accuracy on this database is 99.2% by the $ZM_{component}LTP$ approach [33]. The results indicate that the implemented approaches show robustness to poses (up to $\pm 20°$), slight occlusion, scale, position, and tilt variations as well.

Comparing the results of the proposed approach and the state-of-the art approaches on the Yale database (Table 7), it is observed that the achieved highest recognition rate is 99.5%, whereas the literature reports 100% accuracy for the Block-based S-P approach [13]. It is worth mentioning here that the Block-based S-P approach has been tested only on a single randomly generated dataset. In the present study, results have been illustrated as means of several possible simulations of randomly generated training and test sets. The comparative analysis, therefore, ascertains the robustness of the developed approaches against illumination variations.

Table 6 Performance comparison with state-of-the-art methods on the ORL database

Method	Recognition rate (%)
Linear Discriminant Analysis (LDA, 2008) [51]	91
Direct LDA (D-LDA, 2008) [51]	92.5
2D-LDA (2008) [51]	92.5
Two Dimensional LDA (2D-WLDA, 2008) [51]	93.5
2D-DWLDA (2008) [51]	94
Combined Global and Local Preserving Feature (CGLPF, 2010) [34]	91.39
Complex Zernike Moments (CZM, 2011) [31]	96.5
Complex Wavelet Moments, CWM (2013) [30]	96
Intrinsicfaces (2010) [40]	97
DCT + EFM (Combined Discrete Cosine Transform and Enhanced Fisher Linear Discriminant Model, 2004) [52]	93.2
Feature Fisher classifier (F3C, 2004) [52]	94.9
Combined Feature Fisher classifier (CF2C, 2006) [53]	96.8
Block-based S-P (2011) [13] (on single random dataset)	99
GZMs+dwpLWLD (Combined ZM and Weber Law Descriptor, 2012) [32]	98
ZM_{mag}LTP (Magnitudes of ZM with Local Ternary Pattern, 2014) [33]	98.85
$ZM_{component}$LTP (Real and imaginary components of ZM with Local Ternary Pattern, 2014) [33]	99.2
Two-directional two-dimensional Principal Component Analysis (2D2PCA, 2010) [21]	90.5
Wavelet + LDA (2015) [22]	97.1
Local Directional Pattern (LDP, 2014) [29]	91
Two-dimensional PCA with LDA (2DPCA + LDA, 2014) [54]	91.2
OC-LBP+HOG - GA-SVM	***99.5***
GFZM-GA-SVM	***99.8***

From Table 8 with comparison to the FERET gallery and probe sets, it is noted that on the results for the fa and fc datasets, the LLGP and LGBP methods presented in the literature have the highest recognition rates of 99% and 99.6% respectively. The proposed OC-LBP+HOG approach achieves a recognition accuracy of 99.7%.

This technique achieves improved results especially on the dup I as well as the dup II datasets. It is well-known that these two datasets are extremely challenging for face recognition approaches as they not only exhibit aging variations but also reflect the changes in background and clothing. Thus, the integrated framework of fusing OC-LBP and HOG features is more robust to extreme expression and aging variations. Further, these descriptors are computationally very fast which makes them highly feasible in all types of conditions.

Table 7 Performance comparison with state-of-the-art methods on the Yale database

Method	Recognition rate (%)
Linear Discriminant Analysis (LDA, 2008) [51]	81.89
Direct LDA (D-LDA, 2008) [51]	93.2
2D-LDA (2008) [51]	86.57
Two Dimensional LDA (2D-WLDA, 2008) [51]	88
2D-DWLDA (2008) [51]	89.33
Intrinsicfaces (2010) [40]	74
DCT + EFM (Combined Discrete Cosine Transform and Enhanced Fisher Linear Discriminant Model , 2004) [52]	93.9
F3C (2004) [52]	96.4
Combined feature Fisher classifier (CF2C, 2006) [53]	96.9
Gabor + LBP + LPQ (2013) [54]	90.7
GELM (2015) [28]	82.3
Gabor + DSNPE (2012) [27]	93.5
Block-based S-P (2011) [13] (on single random dataset)	100
GZMs+dwpLWLD (2012) [32]	94.11
ZM_{mag}LTP (2014) [33]	97.56
$ZM_{component}$ (2014) [33]	97
OC-LBP+HOG-GA-SVM	*99*
GFZM-GA-SVM	*99.5*

Table 8 Performance comparison with state-of-the-art methods on the FERET gallery and probe sets

Method	fb	fc	dup I	dup II
Combined HOG features at 8×8–28×28 scales using product rule (2011) [12]	95.4	84	74.6	69.2
FHOGC (2014) [38]	98.3	93.3	86.3	81.2
Multi-scale LBP (2007) [6]	98.6	71.1	72.2	47.4
LBP (2004) [1]	97	79	66	64
Boosted LBP (2004) [45]	97.9	–	–	–
HGPP (2007) [47]	97.6	98.9	77.7	76.1
LLGP (2009) [43]	99	99	80	78
LGBPHS (2005) [46]	98	97	74	71
LGBP (2006) [20]	99.6	99	92	88.9
ELGBP (2006) [48]	99	96	78	77
GPCA (2009) [38]	96	84	70	57
Gabor-EBGM (1997) [42]	87.3	38.7	42.8	22.7
PCA (1991) [2]	85.3	65.5	44.3	21.8
HOG-EBGM (2008) [2]	95.5	81.9	60.1	55.6
Gabor + LBP (2007) [37]	98	98	90	85
$ZM_{magPhase}$LBP (2014) [33]	98.1	91.5	69.4	66.5
LDA [38]	72.1	41.8	41.3	15.4
OC-LBP+HOG- GA-SVM	*99.5*	*99.7*	*96.2*	*93.8*
GFZM -GA-SVM	*88.5*	*99.3*	*76.4*	*75.8*

8 Performance Evaluation on the Self-generated Database

To test the performance for pose variations, experiments have been performed with different training and test sets. Figure 16 shows the images of one subject in different pose variations from the self-generated database.

Firstly, the classifier is trained to recognize face images with pose changes up to a maximum of $\pm 10°$. Testing is done to identify face images with poses $\pm 20°$, $\pm 30°$, $\pm 40°$, and $\pm 60°$. In the next experiment, the classifier is trained for features of only frontal face images while testing is carried out for face images with poses $\pm 20°$, $\pm 30°$, and $\pm 45°$. Table 9 depicts the formation of both training and test sets and presents the recognition results of the completed experiments.

It is inferred from the results presented in Table 9 that the OC-LBP+HOG-GA-SVM and GFZM-GA-SVM show almost similar performances in all the cases. However, GFZM exhibits a slight improvement in performance in comparison to the OC-LBP+HOG approach due to their rotation as well as pose invariance capabilities. In the case of a test set comprising images with poses up to $\pm 20°$, the highest recognition accuracy of 99.6% was obtained with the GFZM-GA-SVM method. While testing for poses up to $\pm 30°$, GFZM has shown the highest recognition rate of 92.4%. In the third experiment for pose variations of up to $\pm 45°$, the OC-LBP+HOG method exhibits the highest recognition accuracy of 83.1%.

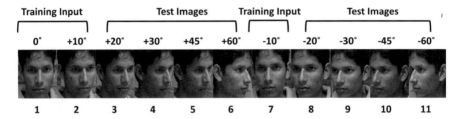

Fig. 16 Face images of one subject in different pose variations ($0° - \pm 90°$) from the self-generated database

Table 9 Experimental results for pose variations on the self-generated database

Experimental datasets		Method	
Training	Testing	OC-LBP+HOG-GA-SVM	GFZM GA-SVM
1,2,7 ($\pm 10°$)	1,2,3,7,8 ($\pm 20°$)	99	99.6
1,2,7 ($\pm 10°$)	1,2,3,4,7,8,9 ($\pm 30°$)	91.1	92.4
1,2,7 ($\pm 10°$)	1,2,3,4,5,7,8,9,10 ($\pm 45°$)	83.1	82
1,2,7 ($\pm 10°$)	1-11 ($\pm 60°$)	74.5	74.6
1 ($0°$)	1,2,7 ($\pm 20°$)	93.3	93.3
1 ($0°$)	1,2,3,7,8 ($\pm 30°$)	84	84
1 ($0°$)	1,2,3,4,7,8,9 ($\pm 45°$)	73.6	75.7
Mean		85.5	85.9

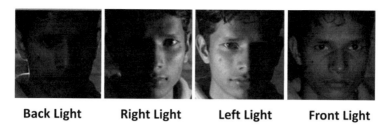

Back Light   **Right Light**   **Left Light**   **Front Light**

Fig. 17 Face images of one subject in different light variations from the self-generated database

Table 10 Recognition accuracy for different lighting conditions on the self-generated database

Method	Training set	
	Random 1 image	Random 2 images
OC-LBP+HOG-GA-SVM	96	98
GFZM-GA-SVM	94.4	97

For images up to $\pm 60°$, both methods present similar results. In the experiments involving only a single training frontal image and rest testing images with poses of up to $\pm 20°$, $\pm 30°$, and $\pm 45°$, both integrated approaches perform equally well and exhibit the highest recognition accuracy of 93.3%, 84.0%, and 75.7% respectively.

Experiments have also been carried out to test the performance of the developed techniques in the case of light variations. Firstly, training is done on one image randomly selected from the images with light changes of each subject as shown in Fig. 17. Testing is done on five images including frontal image with no light effect and the four images with light variations.

In the next experiment, training is done on two randomly selected images and testing on all the five images as considered in the first experiment. The results are shown in Table 10.

Among the results depicted in Table 10, it is observed that the OC-LBP+HOG approach shows the best performance and exhibits up to a 2.4% improvement in recognition accuracy as compared to the GFZM method. As both HOG and OC-LBP techniques are highly effective for light alterations, their combination yields the recognition rates of 96% and 98%. GFZM also achieves reasonably good results.

9 Performance Comparison of the Grid-Based and GA-SVM Model

The study compares the grid-search-based SVM model with the proposed GA-based model. Figures 18 and 19 show the classification accuracy obtained using both the approaches on the different subsets of the three experimental databases using the proposed OC-LBP+HOG and GFZM feature extraction methods. For the ORL and

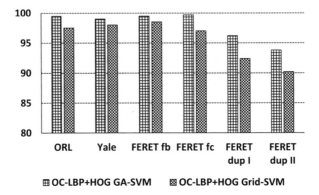

Fig. 18 Comparison of classification accuracy obtained using the GA-SVM model and grid-search-based SVM using the OCLBP+HOG feature extraction method

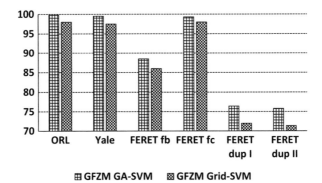

Fig. 19 Comparison of classification accuracy obtained using the GA-SVM model and Grid-search-based SVM using the GFZM feature extraction method

Yale databases, five images per person are used in the training set and testing is done on the remaining five and six images respectively. It is observed that the overall accuracy achieved with the GA-based SVM model is superior to its traditional counterpart. In a nutshell, an improvement of 4.4% is observed in the recognition results using the GA-SVM approach.

9.1 Time Efficiency

Experiments have also been performed to compare the time efficiency of both models to compute the parameters. The results have, hence, been obtained for different k-cross-validation ranges, with k ranging from 3 to 7. Experimentation has been done on the ORL database. As can be seen from Table 11, the time efficiency of the GA-based model is superior to its grid-based counterpart. Taking the fivefold

Table 11 Time requirement for Grid-based and GA-based SVM models

	SVM model	
k	Grid-SVM	GA-SVM
3	812.8	387.1
4	1045.6	612.4
5	1326.7	720.3
6	1891.1	823.4
7	2122.8	941.6

cross-validation for example, the parameter optimization and the classification time for the GA-based model is 720.3 s while that of the grid-based model is 1326.7 s.

10 Conclusion

In the present study, two hybrid feature extraction methods along with a GA and SVM-based hybrid metaheuristic classification model (GA-SVM) have been proposed for face recognition. Owing to the fact that both texture and edge information is required for face representation, the study has inspected one of the hybrid feature extraction methods based on the combination of recently introduced OC-LBP and HOG descriptors. The other method is based on the integration of the Gabor filters and Zernike moments in order to represent faces in varying conditions of pose, orientation, and scale. In the GA-SVM model, a GA-based strategy has been used for the parameter optimization of the SVM classifier. This was done since the selection of parameter values for this classifier is crucial and has a direct influence on the final accuracy achieved. Classification of the developed feature extraction methods was achieved by using different distance-based metrics, including χ^2, square-chord, and extended canberra, as well as by the GA-SVM model with RBF and χ^2 kernels. A comparison of the results of a GA-based parameter optimization strategy to the traditional grid-based algorithm has also been made to depict its applicability. A face database was also created in this study and the performance of the implemented techniques verified against that database. The techniques proposed through this work have been evaluated for varying poses, light conditions, expression changes, occlusion (especially of eyes with eye-glasses), and aging. These have been thoroughly compared with the state-of-the-art face recognition algorithms for establishing their recognition performance. Experimental results indicate the suitability of the proposed recognition frameworks in real-life scenarios.

Acknowledgements The authors are grateful to the National Institute of Standards and Technology, AT&T Laboratories, and the Computer Vision Laboratory, Computer Science and Engineering Department, University of California, San Diego for providing the FERET, ORL, and Yale face databases respectively. During the course of this study, the first author was funded by an INSPIRE-Junior Research Fellowship (IF120810) from the Department of Science and Technology, Govt. of India.

References

1. T. Ahonen, A. Hadid, M. Pietikäinen, Face recognition with local binary patterns, in *European Conference on Computer Vision* (Springer, Berlin, 2004), pp. 469–481
2. A. Albiol, D. Monzo, A. Martin, J. Sastre, A. Albiol, Face recognition using hog–ebgm. Pattern Recogn. Lett. **29**(10), 1537–1543 (2008)
3. A. Amine, M. Rziza, D. Aboutajdine, Svm-based face recognition using genetic search for frequency-feature subset selection, in *Image and Signal Processing* (Springer, Berlin, 2008), pp. 321–328
4. P.N. Belhumeur, J.P. Hespanha, D.J. Kriegman, Eigenfaces vs. fisherfaces: recognition using class specific linear projection. IEEE Trans. Pattern Anal. Mach. Intell. **19**(7), 711–720 (1997)
5. A.A. Bhuiyan, C.H. Liu, On face recognition using gabor filters. World Acad. Sci. Eng. Technol. **28**, 51–56 (2007)
6. C.H. Chan, J. Kittler, K. Messer, Multi-scale local binary pattern histograms for face recognition, in *Advances in Biometrics* (2007), pp. 809–818
7. C.C. Chang, C.J. Lin, Libsvm: a library for support vector machines. ACM Trans. Intell. Syst. Technol. **2**(3), 27 (2011)
8. J.S. Chou, M.Y. Cheng, Y.W. Wu, A.D. Pham, Optimizing parameters of support vector machine using fast messy genetic algorithm for dispute classification. Expert Syst. Appl. **41**(8), 3955–3964 (2014)
9. I.J. Cox, J. Ghosn, P.N. Yianilos, Feature-based face recognition using mixture-distance, in *1996 IEEE Computer Society Conference on Computer Vision and Pattern Recognition, 1996. Proceedings CVPR'96* (IEEE, Washington, 1996), pp. 209–216
10. N. Dalal, B. Triggs, Histograms of oriented gradients for human detection, in *IEEE Computer Society Conference on Computer Vision and Pattern Recognition, 2005. CVPR 2005*, vol. 1 (IEEE, Washington, 2005), pp. 886–893
11. O. Déniz, M. Castrillon, M. Hernández, Face recognition using independent component analysis and support vector machines. Pattern Recogn. Lett. **24**(13), 2153–2157 (2003)
12. O. Déniz, G. Bueno, J. Salido, F. De la Torre, Face recognition using histograms of oriented gradients. Pattern Recogn. Lett. **32**(12), 1598–1603 (2011)
13. M. El Aroussi, M. El Hassouni, S. Ghouzali, M. Rziza, D. Aboutajdine, Local appearance based face recognition method using block based steerable pyramid transform. Signal Process. **91**(1), 38–50 (2011)
14. N.H. Foon, Y.H. Pang, A.T.B. Jin, D.N.C. Ling, An efficient method for human face recognition using wavelet transform and Zernike moments, in *Proceedings. International Conference on Computer Graphics, Imaging and Visualization, 2004. CGIV 2004* (IEEE, Washington, 2004), pp. 65–69
15. J.B. Gao, Q.H. Cao, Adaptive hog-lbp based learning for palm tracking. Adv. Mater. Res. **756**, 3707–3711 (2013)
16. M. Ghorbani, A.T. Targhi, M.M. Dehshibi, Hog and lbp: towards a robust face recognition system, in *2015 Tenth International Conference on Digital Information Management (ICDIM)* (IEEE, Washington, 2015), pp. 138–141
17. Z.M. Hafed, M.D. Levine, Face recognition using the discrete cosine transform. Int. J. Comput. Vis. **43**(3), 167–188 (2001)
18. M. Heikkilä, M. Pietikäinen, C. Schmid, Description of interest regions with local binary patterns. Pattern Recogn. **42**(3), 425–436 (2009)
19. G. Huang, Fusion (2d) 2 pcalda: a new method for face recognition. Appl. Math. Comput. **216**(11), 3195–3199 (2010)
20. D. Huang, C. Shan, M. Ardabilian, Y. Wang, L. Chen, Local binary patterns and its application to facial image analysis: a survey. IEEE Trans. Syst. Man Cybern. Part C Appl. Rev. **41**(6), 765–781 (2011)
21. P. Huang, C. Chen, Z. Tang, Z. Yang, Feature extraction using local structure preserving discriminant analysis. Neurocomputing **140**, 104–113 (2014)

22. Z.H. Huang, W.J. Li, J. Wang, T. Zhang, Face recognition based on pixel-level and feature-level fusion of the top-level's wavelet sub-bands. Inform. Fusion **22**, 95–104 (2015)
23. M. Lades, J.C. Vorbruggen, J. Buhmann, J. Lange, C. von der Malsburg, R.P. Wurtz, W. Konen, Distortion invariant object recognition in the dynamic link architecture. IEEE Trans. Comput. **42**(3), 300–311 (1993)
24. M. Li, X. Zhou, X. Wang, B. Wu, Genetic algorithm optimized svm in object-based classification of quickbird imagery, in *2011 IEEE International Conference on Spatial Data Mining and Geographical Knowledge Services (ICSDM)* (IEEE, Washington, 2011), pp. 348–352
25. Z. Liu, C. Liu, Fusion of color, local spatial and global frequency information for face recognition. Pattern Recogn. **43**(8), 2882–2890 (2010)
26. G.H. Liu, J.Y. Yang, Content-based image retrieval using color difference histogram. Pattern Recogn. **46**(1), 188–198 (2013)
27. G.F. Lu, Z. Jin, J. Zou, Face recognition using discriminant sparsity neighborhood preserving embedding. Knowl.-Based Syst. **31**, 119–127 (2012)
28. Y. Peng, S. Wang, X. Long, B.L. Lu, Discriminative graph regularized extreme learning machine and its application to face recognition. Neurocomputing **149**, 340–353 (2015)
29. H. Ren, H. Ji, Nonparametric subspace analysis fused to 2dpca for face recognition. Optik—Int. J. Light Electron. Opt. **125**(8), 1922–1925 (2014)
30. C. Singh, A.M. Sahan, Face recognition using complex wavelet moments. Opt. Laser Technol. **47**, 256–267 (2013)
31. C. Singh, N. Mittal, E. Walia, Face recognition using Zernike and complex Zernike moment features. Pattern Recognit. Image Anal. **21**(1), 71–81 (2011)
32. C. Singh, E. Walia, N. Mittal, Robust two-stage face recognition approach using global and local features. Vis. Comput. **28**(11), 1085–1098 (2012)
33. C. Singh, N. Mittal, E. Walia, Complementary feature sets for optimal face recognition. EURASIP J. Image Video Process. **2014**(1), 35 (2014)
34. K.R. Soundar, K. Murugesan, Preserving global and local information–a combined approach for recognising face images. IET Comput. Vis. **4**(3), 173–182 (2010)
35. H. Spies, I. Ricketts, Face recognition in Fourier space, in *Vision Interface*, vol. 2000 (2000), pp. 38–44
36. V. Štruc, R. Gajšek, N. Pavešic, Principal gabor filters for face recognition, in *Proceedings of the 3rd IEEE International Conference on Biometrics: Theory, Applications and Systems* (IEEE Press, Piscataway, 2009), pp. 113–118
37. X. Tan, B. Triggs, Fusing gabor and lbp feature sets for kernel-based face recognition, in *International Workshop on Analysis and Modeling of Faces and Gestures* (Springer, Berlin, 2007), pp. 235–249
38. H. Tan, B. Yang, Z. Ma, Face recognition based on the fusion of global and local hog features of face images. IET Comput. Vis. **8**(3), 224–234 (2013)
39. M. Turk, A. Pentland, Eigenfaces for recognition. J. Cogn. Neurosci. **3**(1), 71–86 (1991)
40. Y. Wang, Y. Wu, Face recognition using intrinsicfaces. Pattern Recogn. **43**(10), 3580–3590 (2010)
41. X. Wang, T.X. Han, S. Yan, An hog-lbp human detector with partial occlusion handling, in *12th International Conference on Computer Vision, 2009 IEEE* (IEEE, Washington, 2009), pp. 32–39
42. L. Wiskott, N. Krüger, N. Kuiger, C. Von Der Malsburg, Face recognition by elastic bunch graph matching. IEEE Trans. Pattern Anal. Mach. Intell. **19**(7), 775–779 (1997)
43. S. Xie, S. Shan, X. Chen, X. Meng, W. Gao, Learned local gabor patterns for face representation and recognition. Signal Process. **89**(12), 2333–2344 (2009)
44. A.L. Yuille, P.W. Hallinan, D.S. Cohen, Feature extraction from faces using deformable templates. Int. J. Comput. Vis. **8**(2), 99–111 (1992)
45. G. Zhang, X. Huang, S.Z. Li, Y. Wang, X. Wu, Boosting local binary pattern (lbp)-based face recognition, in *Sinobiometrics* (Springer, Berlin, 2004), pp. 179–186

46. W. Zhang, S. Shan, W. Gao, X. Chen, H. Zhang, Local gabor binary pattern histogram sequence (lgbphs): a novel non-statistical model for face representation and recognition, in *Tenth IEEE International Conference on Computer Vision, 2005. ICCV 2005*, vol. 1 (IEEE, Los Alamitos, 2005), pp. 786–791

47. B. Zhang, S. Shan, X. Chen, W. Gao, Histogram of gabor phase patterns (hgpp): a novel object representation approach for face recognition. IEEE Trans. Image Process. **16**(1), 57–68 (2007)

48. W. Zhang, S. Shan, L. Qing, X. Chen, W. Gao, Are gabor phases really useless for face recognition? Pattern Anal. Appl. **12**(3), 301–307 (2009)

49. J. Zhang, K. Huang, Y. Yu, T. Tan, Boosted local structured hog-lbp for object localization, in *2011 IEEE Conference on Computer Vision and Pattern Recognition (CVPR)* (IEEE, Washington, 2011), pp. 1393–1400

50. D. Zhang, J. Xiao, N. Zhou, M. Zheng, X. Luo, H. Jiang, K. Chen, A genetic algorithm based support vector machine model for blood-brain barrier penetration prediction. Biomed. Res. Int. **2015**, 292683 (2015)

51. R. Zhi, Q. Ruan, Two-dimensional direct and weighted linear discriminant analysis for face recognition. Neurocomputing **71**(16), 3607–3611 (2008)

52. D. Zhou, X. Yang, Feature fusion based face recognition using efm, in *International Conference Image Analysis and Recognition* (Springer, Berlin, 2004), pp. 643–650

53. D. Zhou, X. Yang, N. Peng, Y. Wang, Improved-lda based face recognition using both facial global and local information. Pattern Recogn. Lett. **27**(6), 536–543 (2006)

54. S.R. Zhou, J.P. Yin, J.M. Zhang, Local binary pattern (lbp) and local phase quantization (lbq) based on gabor filter for face representation. Neurocomputing **116**, 260–264 (2013)

55. C. Zhu, C.E. Bichot, L. Chen, Image region description using orthogonal combination of local binary patterns enhanced with color information. Pattern Recogn. **46**(7), 1949–1963 (2013)

Optimization of a HMM-Based Hand Gesture Recognition System Using a Hybrid Cuckoo Search Algorithm

K. Martin Sagayam, D. Jude Hemanth, X. Ajay Vasanth, Lawerence E. Henesy, and Chiung Ching Ho

Abstract The authors develop an advanced hand motion recognition system for virtual reality applications using a well defined stochastic mathematical approach. Hand gesture is a natural way of interaction with a computer by interpreting the primitive characteristics of gesture movement to the system. This concerns three basic issues: (1) there is no physical contact between the user and the system, (2) the rotation of the hand gesture can be determined by the geometric features, and (3) the model parameter must be optimized to improve measurement of performance. A comparative analysis of other classification techniques used in hand gesture recognition is carried out on the proposed work hybrid with the bio-inspired metaheuristic approach, namely the cuckoo search algorithm, for reducing the complex trajectory in the hidden Markov model (HMM) model. An experimental result is as to how to validate the HMM model, based on the cost value of the optimizer, in order to improve the performance measures of the system.

Keywords Virtual reality · Stochastic mathematical approach · Shape-based features · Gesture recognition · Cuckoo search algorithm

1 Introduction

In the present scenario, human–computer interaction (HCI) has made a tremendous change in our society based on gesture recognition for the development of a virtual

K. M. Sagayam (✉) · D. J. Hemanth · X. A. Vasanth
School of Engineering and Technology, Karunya University, Coimbatore, India

L. E. Henesy
Department of Systems and Software Engineering, Blekinge Institute of Technology, Karlskrona, Sweden
e-mail: larry.henesey@bth.se

C. C. Ho
Department of Computing and Informatics, Multimedia University, Cyberjaya, Malaysia
e-mail: ccho@mmu.edu.my

© Springer International Publishing AG, part of Springer Nature 2018
S. Bhattacharyya (ed.), *Hybrid Metaheuristics for Image Analysis*,
https://doi.org/10.1007/978-3-319-77625-5_4

reality system. Human hand gesture is a non-verbal communication, an instinctive, creative, and natural relating to a computer. The main objective of the system is to recognize the hand gesture by determining hand gesture features such as shape, orientation, and velocity, which can be segmented from the static background effect. In this modern age, electronic gadgets are mostly interacted with by touch. This technology has predominately supported the evolution of the touchless device. Most of the real time applications are based on hand gesture control without any physical contact between the computer and the human [38]. In the last decade, little attention has been given to HCI in the field of pattern recognition, such as movement of body, speech, voice, or hand gesture, which can be enriched by a natural user interface with the system. In virtual reality applications, HCI has been very significant in developing a system with the gesture patterns of the whole human body. There are a few artifacts which interact with the system, which was deployed by the Kinect sensor developed by Microsoft Ltd. When considering the entire body movement, the system has to be predefined by the shapes of the human body, like fingers, noses, eyes, hands, lips, and legs. These small dimensional objects are incorporated with the Kinect sensors for determining the correct decisions based on each action given by the users. In this work, an extension of the Kinect sensor based on contour analysis has been adopted to the system for locating the position of the human gesture. In addition to the Kinect sensor output, Graphics Processing Unit (GPU)-based gesture recognition in real time using a Leap Motion sensor has been compared with the various strategies of natural interaction towards the system. This technique directly maps onto two-dimensional (2D) windows, icons, menus, and pointers that convert into a three-dimensional (3D) user interface in the 3D environment [8]. The primary constraint of hand gesture recognition in a 2D environment is the loss of information in any one of the coordinate axes (z-axis). This leads sometimes to a wrong interpretation of the system. Nowadays, most electronic devices are inculcated with gesture recognition in a 3D environment for reducing the computational complexity in the touch devices. A few options in touch devices, like forward, backward, left, and right buttons, are replaced by sensors in touchless devices. In this work we propose a stereo vision based approach for classifying the centroid movement and intensity level of pixels using a conditional random field. This provides the depth image from the acquired input while calibrating the device. The system is robust towards Arabic numerals from 0 to 9, color-based segmentation, and variation of the intensity level of pixels [26]. Based on previous work, [30] an approach is proposed for developing an intangible interface with the help of computer vision technology for HCI using dynamic hand gestures. In the desktop or laptop system, the input device such as a mouse or joystick has basic control like left, right, and a scroll button. It can be controlled by hand gestures in different illumination conditions. After acquisition of real data, it segments the hand portion and converts it into a binary value which is then passed to the system for processing. It should be kept in mind that performance measures like recognition rate and accuracy are the most important values in the design of the hand gesture recognition system based on a stochastic mathematical model. Hand gestures can be used to control the system virtually, as in games, media players, and education. The detailed case study about hand posture and gesture recognition based on a different machine learning algorithm for virtual reality applications has been discussed. The number of potential applications based on hand gestures used in various fields should be acknowledged [35].

1.1 Literature Survey on a Hand Gesture Recognition System Based on HMM

In recent years, most technologies are emerging based on touchless electronic gadgets. Sign language recognition is the most predominant factor in HCI. It can be helpful for recognizing English alphabets by deaf and dumb people by continuous gesture patterns using an artificial neural network [4]. While hand gesture movement is continuous in a 3D environment, it is hard for the system to recognize the critical conditions that occur in a moment. The continuous gesture pattern divides into 'n' number of frames without any artifacts using a spotting algorithm. The performance of the system is improved by efficiency and accuracy [55]. This article has focused on high dimensionality and the redundant feature value of gesture patterns stored in the code book of a HMM model. It can be compressed without change of representation using sparse coding (SC). A hybrid technique hidden Markov model (HMM) +SC and vector quantization has been applied to generate the effective utilization of the data in the code book of the HMM model [34]. A new method has been proposed [12] to control robots by using gesture recognition using HMM. This work is mainly focused on the mapping of human gestures to robots in a 3D environment. Human gestures seamlessly provide normal action to the system, in which the commands are properly transmitted to the robots. This can be perform by mounting an Red-Green-Blue-Depth (RGB-D) camera on the robot to acquire 3D data, and then simultaneously transmitting data to the remote server, finally rendering it into a Virtual Reality (VR) device. On the other hand, the intention of the user is inferred using the recognition process of the motion of head movement which is again made purely on the basis of HMMs. This is later interpreted into commands in order to control the robot. Based on the HMM a dynamic hand gesture interface is introduced for the virtual environment. This model is employed to represent the continuous dynamic gestures where its parameters are learned from the training data which is collected from the cyber glove. Gesture spotting is a major problem faced and thus to avoid it standard deviation is employed for measuring the variation in an angle for each joint in a finger print which uniquely describes the character of gestures. A prototype is applied to the three dynamic gestures which control the rotational directions of a 3D cube that has been implemented to test the effectiveness of the model that is proposed [9].

Apart from all this, the work focused on the evaluation and development of the haptic enhanced system which is virtually real. This system amazingly allows a user to make a handshake with a virtual partner by the use of this haptic interface.

Feedback signals that are multi-modal in nature are designed to generate the illusion which improvises the feel of making a handshake with another human, but in reality it is a robotic arm. An advanced controller from the interfacing field is developed which ultimately responds to the user's behavior in online mode. HMM approaches to human interaction and strategy estimation are the techniques used to achieve online behavior. In order to evaluate the performance of the system, human–robot handshake experiments were carried out. The haptic rendering was compared by means of two different approaches. One was by introducing a basic controller in normal mode along with an embedded curve in the robot which disregards the human partner. The later approach was an introduction of an interactive robot controller for online behavior estimation. These two approaches were matched with the partner behavior of another human driving the robot via tele-operation rather than implementing a virtual partner. In the final estimation, the interactive controller is used to recognize the human using basic controller mode. This concept concentrates on the development of the haptic rendering approach for a hand shaking system. The subjective analysis of an experiment was integrated with visual and haptic cue reports which were analyzed [47]. This work was extended to acknowledge 3D positions, device information which is shared, and also the input from hand gestures. The model was designed by the discrete HMM which were basically rated around 80%; the complex ones amounted to 60% approximately [11]. A low cost acquisition system was adopted for hand gesture recognition on the basis of the various conditions. The possible sets of gesture patterns are trained and stored in the C library for implementing in real time applications [12].

1.2 Literature Survey on the Optimization of the Hidden Markov Model

In order to reduce recursiveness and self looping problem in HMM, a new approach was developed using swam based optimization technique (Fig. 1). An optimization approach is to find the maxima or minima values based on the objective function defined in the system model. This can be an iterative process to achieve better cost value of the given data to the model [39]. The optimization problem deals with managing nature-inspired concepts by choosing the best option in the sense of the given target work. It is broadly classified into two types: the metaheuristic and the heuristic approach. The heuristic approach is a problem-designed rule that has its own decisions for certain issues. The metaheuristic approach is a problem solver that can be finely tuned for various optimization problems by legitimately adjusting the parameters of the system. It can be further classified into three different types: swarm-based, trajectory-based, and evolutionary-algorithm (EA) based optimization [5]. In a machine learning algorithm, nature-inspired algorithms

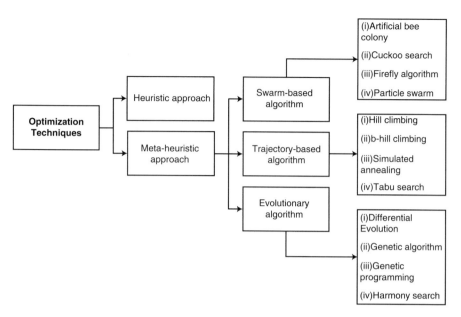

Fig. 1 Various types of optimization techniques

are very essential to get the optimum value for better convergence within a limited run time. This approach narrates the behavior of a group of animals searching for their food with less significant time, climatic condition, and duration. The Krill-Herd algorithm can be used in any research field as a hybridization, multi-objective, and parameter-less function [6]. Based on the different searching methods, a few bio-inspired characters are recognized in the field of pattern recognition, such as the cuckoo search algorithm (CSA) [53], the firefly algorithm [49], the particle swarm optimization [19], and the artificial bee colony (ABC) [21]. Trajectory-based algorithms begin with a solitary temporary arrangement. At every step, the arrangement will be moved to its neighboring arrangement, with respect to a particular neighborhood structure. Based on the search space in a region, it is classified into four types: β-hill climbing [1], hill climbing [25], simulated annealing (SA) [23], and tabu search [16]. An evolutionary-based algorithm begins with an arrangement of creatures called a population. In every era, EA approaches recombine the ideal qualities of the present generation to transform them into a new generation based on natural methods. Nature inspired selection principle, is classified into four types: harmony search [15], differential evolution [42], genetic programming [24], and genetic algorithm (GA) [18]. The conventional optimization has a few limitations while hybridizing with any machine learning algorithm. In this work, it was proposed that during the training process the GA hybridizes with HMM, which is more optimal than training HMM using the Baum–Welch (BW) approach, as it takes more time for training the HMM model to get higher precision, and recalls the value for web information extraction [48]. The same method has been used for text information selection. During the training phase, the genetic operator combines with the Baum–Welch algorithm to optimize the model parameter of HMM. From the reconstructed HMM, the text information is extracted using the Viterbi algorithm with an optimal state sequence of tested values of text data. GA-HMM is superior to the traditional method of HMM in terms of higher robustness and wide optimization range, whereas it has a defective premature convergence rate [27]. Baum's re-estimation algorithm is used to decode the state sequence in HMM model for getting constant directly for single channel kinetics. This approach is unable to optimize with the constant rate for a single channel directly. A most likelihood function was derived using the quasi-Newton method to get the most efficient optimum value to improve the performance rate by hybridizing the HMM model with the direct optimization approach [31]. The Viterbi algorithm is typically difficult to get to the optimum state sequence in all its observations of the HMM model. An extended version of the Viterbi algorithm optimizes the HMM in real time for multi-target tracking [2]. Any state sequence in the HMM model has a quite complex structure for any problem and is often of higher dimensionality. A complex framework for ancestral population genomics was built. This produced a high prominent result for parameter estimation using the heuristic based optimization approach than the gradient based optimization approach [10].

1.3 Outline

This chapter is organized as follows: Section 2 explains the motivation, problem statement, and framework of the proposed system. Section 3 presents the sample hand dataset and image pre-processing procedure. Section 4 discusses how to extract the hand gesture feature points. Section 5 describes in detail the complexity of the HMM model based on the state sequence of hand gestures. Section 6 presents how to get the optimum value using the CSA. Section 7 shows an experimental result with an analytical discussion. Finally, the conclusion and summary with future enhancement of this research work is shown in Sect. 8.

2 Motivation and Problem Statement

A recent innovation for developing virtual reality applications using hand gesture recognition is listed in Table 1 [35]. Based on the issues, it is proposed to enhance the newer model with a stochastic mathematical approach for Human Computer Interaction (HCI). Glove-based (GB) hand gesture normally wears the glove on the hand and is connected to the system. The interaction is made with the system by folding the hand or fingers. This leads to short circuitry, which affects the robustness of the system. Vision-based (VB) gesture recognition delays recognizing hand movement by the system. The factors affecting this scheme are: (1) the quality

Table 1 Problems in conventional techniques of hand gesture recognition systems [35]

S. no	Techniques	Problem
1	Active shape model + VB	Track only the hand movement
2	Feature extraction + GB	Computational complexity is quite high
3	Finite state machine	Robustness is quite low in rigid condition
4	Linear fingertip model + VB	Not applicable real-time purpose
5	Principal component analysis + VB and GB	Time consumption is significantly more in training process
6	Template matching + VB and GB	Not applicable for huge dataset
7	Time delay neural network	Robustness is quite low in typical input

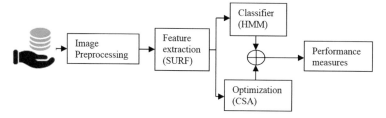

Fig. 2 Framework of hand gesture recognition with the proposed system

of the acquisition device, (2) the recognition rate and accuracy, (3) the user's environment, (4) the usage of a low-level feature point of the given data, and (5) discrepancies of the input data to the system. The conventional methods of hand gesture recognition are replaced by the machine learning approaches, which can learn from the trained data and make predictions to the system. A model can be constructed from the sample data, using statistically programmed instructions by using data-driven decisions. Some of the algorithms can support any data-driven problem such as a support vector machine, a logistic regression, a decision tree, an HMM, a linear regression, or Naive Bayes, K-mean clustering. In most of the research problems, HMM has been used for pattern recognition applications. A few problems in the HMM model are scalability, recursiveness, storage capabilities, fitness value, and unwanted transitions [36, 37]. These problems in the machine learning algorithm can be significantly overcome either by heuristic or metaheuristic optimization approaches. This can reduce the system and computational complexity in the model [5].

2.1 Proposed Work

The framework of the hand gesture recognition system is shown in Fig. 2. In this work, the four major functional blocks are pre-processing, feature extraction, classification, and optimization. First, create the hand gesture dataset of five different classes with 60 frames per second. So, in total 300 hand images are stored in the image dataset, which has been taken from the Cambridge hand dataset. To locate the exact boundary of the hand image from the dataset we have to apply the edge detection technique using its kernel function for determining angular position in free space. To define the hand gesture with the shape-based descriptor which is used to locate the point of interest and converts the target point into coordinate using speeded up robust feature (SURF). After achieving the prescribed feature point from the hand gesture data, train and test using the HMM approach. Basically, the HMM structure, which is constructed with the finite state machine , presents some difficulty when using the dynamic programming approach in finding the optimum path from source to the target point in the model. In order to enhance the fitness value of the HMM model, a new hybrid metaheuristic approach called the CSA is used. This algorithm is used to reduce an unwanted transition state sequence in the HMM model during the training process. This can improve performance measures of the system.

3 Image Database and Pre-processing

This work is carried out with five different hand gestures in the database: flat to left, flat to right, flat to contract, V-shape left, and V-shape right, which are taken from the Cambridge hand dataset from Imperial college, London [36, 37]. Each class has 60 frames per second, dimensions of 320×240, and total 300 hand gestures, which are stored in the database shown in Fig. 3. Consider the case of a hand moving from flat to left position which takes the sample data of the 0th, 15th, 30th, 45th, and 60th

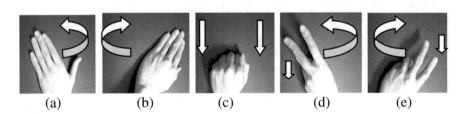

(a)   (b)   (c)   (d)   (e)

Fig. 3 Cambridge hand dataset: five different classes of hand gesture data (**a**) Flat to left (**b**) Flat to right (**c**) Flat to contract (**d**) V-shape to left (**e**) V-shape to right

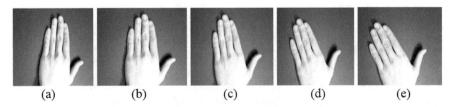

(a)   (b)   (c)   (d)   (e)

Fig. 4 Hand movement from flat to left position at the (**a**) 0th (**b**) 15th (**c**) 30th (**d**) 45th (**e**) 60th frame

-1	-2	-1
0	0	0
1	2	1

-1	0	1
-2	0	2
-1	0	1

Fig. 5 Sobel operator

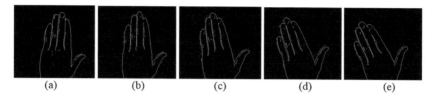

| (a) | (b) | (c) | (d) | (e) |

Fig. 6 Pre-processed hand gestures (**a**) 0th (**b**) 15th (**c**) 30th (**d**) 45th (**e**) 60th frame

frame from the total frames per second as shown in Fig. 4. An input hand gesture pattern shows as a shadow on the screen, which leads to the wrong decision while cropping the edge of the hand data. There are various pre-processing techniques, such as edge detection, smoothing filters, histogram equalization, thresholding, and desaturation. In order to plot and identify the discontinuity in the hand gesture data, the edge detector based on the Sobel operator [17] is used as shown in Fig. 5. Thus the main agenda of pattern recognition is done by the detector. This does the work of cropping the boundary of the hand gesture data with respect to the adjacent background [28]. The pre-processed hand gestures use the coefficient values of the Sobel operator Fig. 5; the sample frames per second are listed at the 0th, 15th, 30th, 45th, and 60th frame, as shown in Fig. 6.

4 Feature Extraction

Machine learning comprises the concept of extracting a feature, which has a large set of data from an original information into a smaller set. Three feature detection methods are basically used for rotational features: (1) a scale invariant feature transform (SIFT), (2) principal component analysis-SIFT (PCA-SIFT), and (3) a SURF transform [20]. Here, in order to extract from the scaling, hand rotation, noise, illumination condition, and direction, the SURF detector is widely used. The over-emphasizing factor of SURF is that it works much faster based on the interest point with less dimensionality [3]. To spot the required relevant point from the data, a Hessian matrix $H(X, \sigma)$ is used.

$$H(X, \sigma) = \begin{bmatrix} L_{xx}(X, \sigma) & L_{xy}(X, \sigma) \\ L_{yx}(X, \sigma) & L_{yy}(X, \sigma) \end{bmatrix} \tag{1}$$

where L represents the Gaussian second-order derivative of the convoluted image at X with scaling of σ.

4.1 Point of Interest

This approach is made purely on the basis of Hessian matrix approximation. To determine the point of interest used by one of the blob detectors a method called SURF based on a Hessian matrix is used. The notation of an integral image $I_\Sigma \mid_{X=(x,y)^T}$ implies the summation of a pixel in the given input data. This allows the performing at a higher rate of an input image which has a rectangular boundary of input I.

$$I_\Sigma(X) = \sum_i \sum_j I(i, j); 0 < i \leq x; 0 < j \leq y \tag{2}$$

The discriminate value of a hand gesture can extract the maximum and minimum from the descriptive features. A feature vector point can be detected from the description of the input data. This initiates the building of the window surrounded by that feature point. Feature points can be calculated from the orientation of a window in a pixel region.

4.2 Descriptor

The surrounding point distributes maximum intensity to the neighborhood and mainly uses a scale space interpolation method to identify the exact interest points. Along the x and y directions, the circular neighborhood response is extracted from a 2D Haar wavelet. This works in such a manner that it slides $\pi/3$ times over and over again until the window size is in a circular manner for every iteration it accesses. Now, based on its orientation it is split into a smaller region which is from a quadratic grid to a square region on the basis of the point of interest as shown in Fig. 7. This split up happens in order to improve the deformation and also the robustness against localized errors on the basis of the response of dx and dy. Thus the relative sum of the oriented points is given by $V = (\Sigma dx, \Sigma dy, \Sigma |dx|, \Sigma |dy|)$; where the resultant value describes the feature point in the vector and, if the vector is unity, the contrast effect is achieved, which is on the basis of variance in a scaling factor [3].

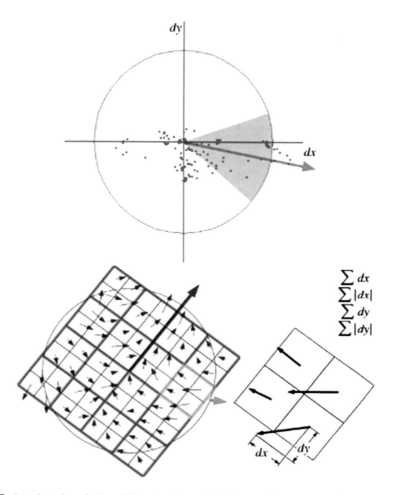

Fig. 7 An oriented quadratic grid based on the point of interest with a square region

4.3 Matching

The orientation of the Laplacian operator is prominently used based on need. It mainly highlights the fact that the feature points are formed perfectly matched; it also ensures that the gestures move at a faster rate. Thus, we can conclude that the SURF algorithm is comparatively more accurate than the SIFT, though both are invariant to rotation, which is applicable for pattern recognition [43].

5   Classification

Pattern recognition plays a vital role in various fields such as signal processing, machine learning, probability, computational geometry, and statistics. It has a significant role in computer vision and artificial intelligence, and for applications in science, engineering, business, and medicine. In the last couple of decades, most of the researchers focused on pattern recognition problems using machine learning algorithms [41]. Classification is a method for arranging pixels and assigning them to particular classes. There are two broad categories of learning scheme: supervised and unsupervised learning. If the learning scheme is said to be supervised then the system has to guide by providing the trained data for each class, whereas unsupervised learning does not provide trained data to the system. In this work, supervised learning was chosen for training and testing the given problem. Here the background of the hand gesture is fixed and modeled using the HMM from the extracted feature point from the contours of hand data [44].

5.1   Hidden Markov Model

The HMM is a stochastic mathematical model for any pattern recognition scheme. It is characterized with 'n' number of states, where 's' denotes an individual state, 'o' denotes the number of different observation variables by states, v denotes an individual symbol and a set initial parameter $\lambda = (A, B, \pi)$, where $A = (a_{ij})$ is the transition state probability matrix with,

$$a_{ij} = P[q_{t+1} = s_j | q_t = s_i], 1 \leq i, j \leq n \tag{3}$$

$B = b_j(k)$ is the distribution probability of the observation symbol.

$$b_j(k) = P[v_{k,t} | q_t = s_j], 1 \leq j \leq n; 1 \leq k \leq 0 \tag{4}$$

and $\Pi = (\pi_i)$ is the distribution state probability of the initial state sequence.

$$\pi_i = P[q_1 = s_i], 1 \leq i \leq n \tag{5}$$

where $s = (s_i), i = 1, 2, \ldots n$ is the state of subsets, qt is a state sequence at each time interval t, and $v = (v_i), i = 1, 2, \ldots n$ is the set of symbols in the state sequence. Three fundamental problems in the HMM model are evaluation, decoding, and training. The evaluation process can be done by calculating the output observable state sequence 'o' with respect to 'λ'. This can be solved by a forward-backward algorithm [14]. The decoding process can be done by determining an optimum state variable which is involved with observable state variable 'o' in the HMM model parameter 'λ'. This can further be solved by a Viterbi algorithm [32].

The training phase is solved by the BW approach, which maximizes the output probability of an observation sequence 'o' [14]. Consider the observation sequence $Q = (Q_1, Q_2, \ldots, Q_T)$ and its model parameter $\lambda = (A, B, \pi)$.

Problem 1 The effectiveness problem to find the probability of observation state sequence $P(Q|\lambda)$.

Problem 2 To set the model parameter $\lambda = (A, B, \pi)$ to find the probability of the observation state sequence $P(Q|\lambda)$ is maximum.

The distribution probability of the observation sequence $P[v_{k,t}|q_t = s_i]$ contains only either continuous or discrete variables. If the probability of the observation variable is continuous then,

$$b_i(k) = \int_i^k P[v_{k,t}|q_t = s_i] \tag{6}$$

If the probability of the observation variable is discrete then,

$$b_i(k) = \sum_i^k P[v_{k,t}|q_t = s_i] \tag{7}$$

If the probability of the observation variable is a vector then,

$$b_i(k) = P[v_{k,t}|q_t = s_i] = N(x|\mu, \Sigma) \tag{8}$$

$$N(x|\mu, \Sigma) = \frac{1}{\sqrt{2\pi^d|\Sigma}} e^{\frac{-1}{2}(x-\mu)^T \Sigma^{-1}(x-\mu)} \tag{9}$$

The Gaussian mixture O of the model is more flexible than others and is expressed as,

$$P[v_{x,t}|q_t = s_i] = \sum_{o=1}^{O} P[O_t = o|q_t = s_i] N(x_{m,i}, \Sigma_{m,i}) \tag{10}$$

where O_t represents a hidden layer with its mixture component; the a priori probability of the conditional weight is $P[O_t = o|q_t = s_i] = C_{i,m}$. In this work, create the model for the hand gesture using HMM for each class of variables.

5.2   Solution to Problem 1

To determine the probability of the observation sequence $Q = (Q_i); i = 1, 2, \ldots T$ with its model parameter $\lambda = (A, B, \pi)$ directly deals with all possible state sequences of all observations of length T. The effective procedure instead of performing the mentioned scheme is called a forward-backward approach. Solving Problem 1, we have to know how to recognize success for a trained model with its observation state sequence. The weight values can be updated iteratively as mentioned in Eqs. (11) and (12).

$$\alpha_t(i) \approx P(o_1, \ldots o_T, q_t = s_i) \tag{11}$$

$$\beta_t(i) \approx P(o_{t+1}, \ldots o_T, q_t = s_i) \tag{12}$$

where $\alpha_t(i)$ represents the probability of state model $q_i(t)$ which initiates t elements of observation sequence Q with step t, $\beta_t(i)$ represents the probability of state model $q_i(t)$ which generates the remainder of the target state sequence, that is from $T \leftarrow t + 1$. To solve $\alpha_t(i)$ and $\beta_t(i)$, there are three basic steps using an inductive approach: initialization, induction, and termination.

1. **Initialization:**

$$\alpha_t(i) = \pi_i b_i(o_i), 1 \leq i \leq n \tag{13}$$

$$\beta_t(i) = 1, 1 \leq i \leq n \tag{14}$$

2. **Induction:**

$$\alpha_{t+1}(j) = \sum_{i=1}^{n} \alpha_t(i) \alpha_{ij} b_j(o_{t+1}), 1 \leq t \leq T - 1; 1 \leq j \leq n \tag{15}$$

$$\beta_t(j) = \sum_{i=1}^{n} a_{ij} b_j(o_{t+1} \beta_t(i)), 1 \leq t \leq T - 1; 1 \leq j \leq n \tag{16}$$

3. **Termination:**

$$P[Q|\lambda] = \sum_{i=1}^{n} \alpha_t(i) = \sum_{i=1}^{n} \beta_t(i) = \sum_{i=1}^{n} \beta_t(i) \alpha_t(i) \tag{17}$$

5.3 Solution to Problem 2

The predominant step is to set the model parameters $\lambda(A, B, \pi)$ and to maximize the probability $P(Q|\lambda)$ because there is no optimal solution of model parameter estimation for a given finite observation state sequence. This can be avoided by locally maximizing its model parameter using a BW algorithm. Let us consider the probability at initial state s_i at time t, and state s_j at $t + 1$, given the parameter and observation state sequence, that is,

$$\psi_t(i, j) = P[q_t = i, q_{t+1} = j | Q, \lambda] \qquad (18)$$

Rewrite Eq. (18) by using the forward-backward algorithm,

$$\psi_t(i, j) = \frac{P[q_t = i, q_{t+1} = j | Q, \lambda]}{P[Q|\lambda]}$$

$$\psi_t(i, j) = \frac{\alpha_t(i)\alpha_{ij}b_j(o_{t+1})\beta_{t+1}(j)}{P[Q|\lambda]}$$

$$\psi_t(i, j) = \frac{\alpha_t(i)\beta_{t+1}(j)a_{ij}b_j(o_{t+1})}{\sum_{i=1}^{n}\sum_{j=1}^{n}\alpha_t(i)\beta_{t+1}(j)a_{ij}b_j(o_{t+1})} \qquad (19)$$

Equation (20) shows the probability value being in the state s_i

$$\gamma_t(i) = \sum_{j=1}^{n} \psi_t(i, j) \qquad (20)$$

This is satisfied by Shotton et al. [40] who maximize $O(\lambda, \overline{\lambda})$ over $\overline{\lambda}$ results to a maximum likelihood for estimation of the HMM model by Eq. (22)

$$O(\lambda, \overline{\lambda}) = \sum_{O} P[O|Q, \lambda] log([O, Q|\overline{\lambda}]) \qquad (21)$$

$$\max_{\overline{\lambda}}[O(\lambda, \overline{\lambda})] = P[Q|\overline{\lambda}] \geq P[Q|\lambda] \qquad (22)$$

The model parameter $\lambda = (A, B, \pi)$ can be re-estimated using:

$$\overline{a_{ij}} = \frac{\sum_{t=1}^{T-1} \psi_t(i, j)}{\sum_{t=1}^{T-1} \gamma_t(i)} \qquad (23)$$

$$\overline{b_j} = \frac{\sum_{j,U_k} s_{j,u_k}}{\sum_j s_j} \qquad (24)$$

At time $t = 1$, the expected state sequence s_i has to re-estimate the model parameter by $\overline{\lambda} = (\overline{A}, \overline{B}, \overline{\lambda})$ using Eqs. (23) and (24), through which it determines an observation state sequence based on maximum likelihood value.

5.4   Inference Problem in HMM

The probability of a classification error making an incorrect decision is due to the following reasons.

1. A Viterbi algorithm is expensive, both in terms of memory and computing time, for a sequence of length n. The dynamic programming for finding the best path through a model with s states and e edges takes a memory proportional to s_n and time proportional to e_n as in Eqs. (20) and (22).
2. The forward-backward approach is even more expensive. It needs to be trained on a set of seed sequences and generally requires a larger seed than the simple Markov models. The training involves repeated iterations of the Viterbi algorithm.

6   Cuckoo Search Optimization

Lots of species of birds have common behaviors and features [33]. In general, eggs of different shapes lie by the mother birds and are secure in the nest for the safeness of the chicks [32]. During reproduction time, birds need a specific nest to live in together, the birds that do not build their own nest are referred as breed parasites. These kinds of birds have laid their eggs in other species' nests, in which their young can be taken care of by the host. In this category, the cuckoo is the most representative. The process requires the cuckoo to install its own eggs in the host nest. However, it deliberately matches the egg with the patterns, color, and shape of the eggs already in the host nest, an expertise that requires a high precision rate. An egg-laid time is a predominant factor for selection of the nest, where the other birds laid eggs in another nest [22, 33]. This process will be evident to the cuckoo birds, whose egg will hatch before the other bird's eggs. This emphasizes clearly that the host bird has taken care of all the chicks in the nest. The gain of a feeding opportunity for the cuckoo's chick will be huge and it will live safely until it is grown [50]. Yang et al. [53] propose the optimization technique using CSA. This has been significantly used in all fields of research such as signal processing, data analytics, pattern recognition, and machine learning [13]. It has been specialized to restrict the character of breed parasitic species such as cuckoo birds, with heavy tail so that they can go higher distance with higher levy flight. It could be inspired by the behavior and character of a few birds and fruit flies. In addition, it is found [7, 29] that in real time levy flight moves randomly with less interval of time [45, 46]. This is an efficient metaheuristic based approach which balances the search space problem in terms of a local search and globally [41].

6.1 Hybridization of the CSA with HMM

There are some distinguishing characteristics of cuckoo birds when laying their eggs [54]. Three idealized conditions are to be follow for the search space problem:

1. At a particular time, one egg is laid by the cuckoo and randomly stored in the nest.
2. A high quality nest acquires the best eggs, which will be preserved for future generations.
3. The host bird's nest is fixed, and to identify the cuckoo bird which lay their egg in other bird's nest can be found using probability condition $P_a \in [0, 1]$. In some cases, either the host will leave from the nest or the cuckoo will construct a new nest by itself. Also, the value of probability P_a signifies the change of nest from the host.

Despite these three conditions, the fundamental procedure for constructing with a pseudo-code the cuckoo search algorithm appears in the algorithm below. One of the essential points of interest of CSA is that it uses fewer parameters to control against compared to numerous other search techniques [15, 18, 39]. Table 2 shows the commonly used values, qualities, and parameters. The points of interest are gathered from [53].If the host discovers the egg is another bird's then there is the possible probability of $P_a \in [0, 1]$ for the egg to be thrown out of the nest. So the cuckoo birds build their own nests in order to protect the younger ones, randomly. This is one of the optimum solutions x^{t+1} of the cuckoo birds [51, 52].

$$x^{t+1} = x^t + a \oplus \text{levy}(\lambda) \tag{25}$$

Algorithm 1: Cuckoo search optimization

Step 1   Initialize the set $X = (x_d; d = 1, , D)^T$; define the objective function f(X)
Step 2   Initial population of n host nests has to be generated $X_i (i = 1, , n)$
Step 3   While t < maximum_iteration do
           (t < maximum_generation) or (stop)
           Get a random cuckoo by levy flight
           Evaluate the fitness function F_i
           Choose a nest among n in random
Step 4   If $F_i > F_j$ then
           Replace j by new optimum solution
           End if
Step 5 P_a fraction of host bird identifies the cuckoo's egg and the cuckoo builds a new nest
Step 6   Keep the chicks of the cuckoo which should be safer (best solution);
Step 7   Determine the best cost value
           End While
           Repeat Step 1

Parameters	Commonly used	Qualities (range)	Symbols
Step size	$a = 1$ [28]	$a > 0$	a
Fraction	$P_a = 0.25$ [57]	[0, 1]	P_a
Nest	$N = 1$ [34]	[15, 56]	N

Table 2 Initial parameter setting of CSA [39]

where step size $a > 0$ represents the scaling of interest to this problem. The symbol $\oplus$ denotes the multiplication entry-wise.

$$\text{levy}(\lambda) \approx \nu = t^{-\lambda} \tag{26}$$

Equation (26) provides the best solution for the cuckoo bird in a random manner to survive their young ones in a protective way.

The concept of CSA is hybridized with HMM during the training process. The model parameter of HMM $\lambda = (A, B, \pi)$ has to incorporate CSA to choose the optimum path state sequence in the forward-backward algorithm using the Viterbi approach. This could find the survival of the fitness function from the objective value defined in the problem. Until the most probable likelihood is maximized, repeat Step 3 to get the best cost value. Or else repeat Step 1 to get the optimum HMM as shown in the algorithm below.

Algorithm 2: Hybridization of CSA with HMM

Step 1 Initialize the number of state sequence 's' as an objective function

Step 2 Fix the observation sequence of state 'o' as the host nest

Step 3 While t < maximum_iteration do

   Choose the optimum path in the forward-backward algorithm using the Viterbi approach survival of fitness function

Step 4 If $F_i > F_j$ then

   Optimum solution

   End if

Step 5 Most probable likelihood function produces the best cost value; **Repeat** Step 3

Step 6 Determine the best cost value

   End While

   Repeat Step 1

---

7   Results and Discussion

In this work, five different classes of Cambridge hand gesture data consists of 60 frames with dimensions of 320×240, so in total 300 images are stored in the database. These gestures are labeled as flat-to-left, flat-to-right, flat-to-contract,

V-shape left, and V-shape right, which are shown in Fig. 3. These data are pre-processed by using an edge detection technique. This is used to crop the hand contour region from the image without cropping the shadow and other content in the image dataset. In edge detection techniques, there are four different operators: Sobel, Prewitt, Canny, and Robert. Among these four, the Sobel operator provides the target object in black and white to represent low and high-level pixel quantities in the image which deliberately avoids false detection, as shown in Fig. 6.

7.1 SURF Features

SURF is used to obtain distinctive features from the given dataset, to reduce the dimensionality of feature points in different angular positions, illumination, and shape. The objective of this work is to reduce the dimensionality of the feature with higher robustness and precision using SURF [3, 43]. This follows the procedure based on Hessian matrix $H(X, \sigma)$ with a scaling 'σ' in the range (0–1) [3]. To measure the performance metrics using SURF features it is recommended that the feature vector point uses two parameters, that is recall and precision, as given in Eqs. (27) and (28).

$$\text{Recall} = \frac{\text{retrieval of correct match}}{\text{total no. of correct match}} \tag{27}$$

$$\text{Precision} = \frac{\text{retrieval of corrected match}}{\text{total no. of matches retrieved}} \tag{28}$$

Table 3 shows the performance metrics of SURF features for the hand gesture data. It has various descriptors based on blurred edges, blurred texture, scaled edges, scaled texture, view point edges, view point texture, illumination, JPEG compression, and different dimensions. This can work with a nearest neighbor distance ratio (NNDR) strategy which selects the corresponding point of interest which represents the minimum Euclidean distance below the threshold value.

Table 3 Performance metrics of SURF features (Cambridge hand gesture data)

Descriptors	Dimension (D)	Recall	Precision
SURF-blurred region	30	[0, 0.8]	[0, 0.65]
SURF-blurred texture	32	[0, 0.6]	[0, 0.78]
SURF-scaled edges	32	[0, 0.59]	[0, 0.85]
SURF-scaled texture	32	[0, 0.55]	[0.85, 0.95]
SURF-view point edges	32	[0, 0.5]	[0.2, 0.88]
SURF-view point texture	32	[0, 0.65]	[0.07, 0.75]
SURF-illumination	32	[0, 0.8]	[0.3, 0.7]
SURF-JPEG compression	28	[0, 0.85]	[0, 0.3]
SURF-different dimensions	NNDR	[0, 0.68]	[0, 0.85]

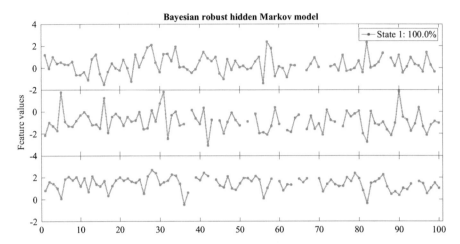

Fig. 8 Estimation of HMM model: feature values vs no. of states $n = 1$

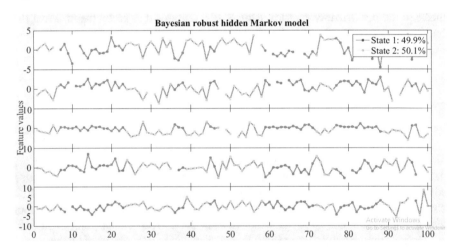

Fig. 9 Estimation of HMM model: feature values vs no. of states $n = 2$

7.2 Classification Results

Among five different classes of hand gesture data defined in the dataset, we have to classify which one belongs to the given target set using HMM. Define the model of HMM parameter $\lambda = (A, B, \pi)$ with respect to the number of states $n = 1, 2, 3$ respectively. Once the trained feature point is passed to the HMM model, a different pattern is shown based on its feature values vs the state sequence as shown in Figs. 8, 9, and 10. For validation of these hand gestures in the HMM model, train first 45 observations of five different classes of hand gesture data using a K-fold cross-validation procedure. Table 4 shows that the training and testing data for

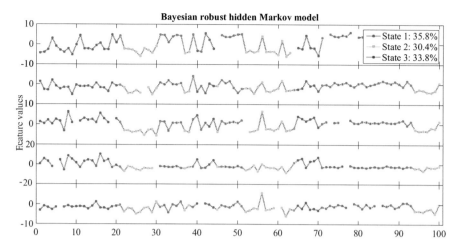

Fig. 10 Estimation of HMM model: feature values vs no. of states $n = 3$

Table 4 Training and testing hand gesture data using cross-validation method

Model	Total no. of frames	K-fold cross-validation	
		Training set	Testing set
Flat-to-left	60	45	25
Flat-to-right	60	40	20
Flat-to-contract	60	48	12
V-shape left	60	42	18
V-shape right	60	46	14

Table 5 Complexity analysis in terms of time and memory

Forward-backward algorithm	Addition time	Multiplication time	Division time	Memory
$\alpha_t(i)$	$s^2t - s_s^2$	$s^2t + st - s^2$	$st - s$	st
$\beta_t(j)$	$s^2t - st - s^2 + 2$	$2s^2t - 2s^2$	st	–
$\sum_{t=1}^{T-1} \gamma_t(i)$	$st - t$	st	st	s
$\sum_{t=1}^{T-1} \psi_t(i,j)$	$s^2t - st - s^2 + s$	$3s^2 - t - 3s^2$	$s^2t - s^2$	s^2
Total	$3s^2t + st - 3s^2 + 3s - t$	$6s^2t + 2st - 6s^2$	$s^2t + 3st - s^2 - s$	$st + s^2 + s$

all classes of data using a K-fold cross-validation method, where K is random integer value based on the different gesture patterns. The internal structure of HMM has a complex construction in relation to the state and observation sequences. It can be derived by a dynamic programming method either by using a forward-backward algorithm or a Viterbi algorithm. The main issue in both algorithms is memory consumption, due to the recursiveness in the model as mentioned in Sect. 5.4. A manual calculation for the forward-backward approach with a Viterbi algorithm is made by an analysis in terms of time and memory as shown in Table 5.

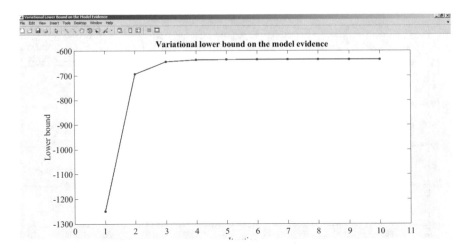

Fig. 11 Lower bounded region based on model evidence w.r.t. no. of states $n = 1$

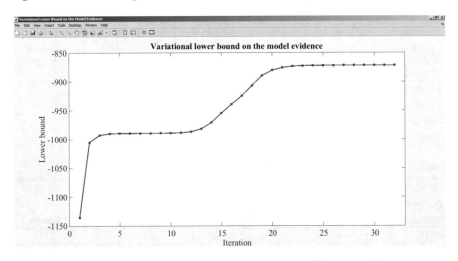

Fig. 12 Lower bounded region based on model evidence w.r.t. no. of states $n = 2$

Figures 11, 12, and 13 show that the lower and upper bound variations based on the model evidence have been presented with respect to the number of states. Table 6 shows the performance result metrics of hand gesture recognition without an optimization technique. This can be validated by using K-fold cross-validation method, which provides an average recognition rate of 80.16% and an error rate of 6.79% respectively.

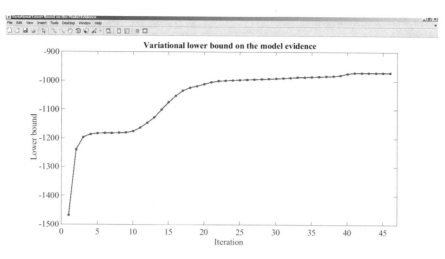

Fig. 13 Lower bounded region based on model evidence w.r.t. no. of states $n = 3$

Table 6 Performance metrics without optimization

Models	Testing data	Recognition rate (%)	Error rate (%)
Flat-to-left	25	85.56	4.35
Flat-to-right	20	75.87	9.52
Flat-to-contract	12	84.45	5.64
V-shape left	18	77.48	6.72
V-shape right	14	77.33	7.72

7.3   Hybridization of HMM with CSA

The model parameters of HMM can be defined in this work as $\lambda = (A, B, \pi)$. It has three parameters to optimize using metaheuristic-based optimization called CSA. Table 7 shows that a different number of state sequences versus cost value is produced by using CSA with various iterations. In addition, the setting up of an initial parameter in CSA is shown in Table 8. Again the system has to be validated with the same procedure as with CSA initial parameters and best cost value. This shows the improved result as listed in Table 9. The testing set of hand gesture data can be determined by HMM incorporated with the CSA algorithm. This produces an average recognition rate of 81.21% and an error rate of 6.17% respectively. Below table shows the comparative analysis of the proposed CSA-HMM with conventional hybridization optimization approach.

Hybridize optimization model	Recognition rate (%)	Error rate (%)
Proposed CSA-HMM	81.21	6.2
ABC-HMM [24]	72.93	3.6
GA-HMM [18]	75.98	4.4

Table 7 Best cost function using CSA

No. of states (n)	Cost value	Iteration
1	1,205,868	4
2	2,280,126.45	4
3	2,827,946.16	3

Table 8 Setting up of an initial parameter in CSA

Initial parameters	Values assigned
No. of cuckoos	1,2,3
Minimum no. of eggs	2
Maximum no. of eggs	4
Maximum iteration	100
KNN cluster numbers	1
Motion coefficient	9
Accuracy	$-$inf
Maximum no. of cuckoos	10
Radius coefficient	5
Cuckoo population variance	$1e^{-13}$

Table 9 Performance metrics with CSA optimization

Models	Testing set	Recognition rate (%)	Error rate (%)
Flat-to-left	25	86.72	4.09
Flat-to-right	20	78.77	8.67
Flat-to-contract	12	84.65	5.02
V-shape left	18	78.02	6.03
V-shape right	14	77.87	7.04

8   Summary and Conclusions

In the present scenario, hand gestures are more significant in the development of virtual reality applications. This work has mainly focused on the optimization of HMM-based hand gesture recognition using a metaheuristic approach known as CSA. The traditional system of HMM for hand gesture recognition has been trained and tested with the proposed method and compared with a GA and an ABC algorithm. A framework of the system consists of four major blocks: pre-processing, feature extraction, classification, and optimization. Firstly, the Cambridge hand gesture dataset stores 300 images with dimensions of 320×240 of five different classes defined as flat-to-left, flat-to-right, flat-to-contract, V-shape left, and V-shape right. These hand gesture images have fixed illumination conditions and each class has 60 frames per second. Secondly, the hand data is cropped from the input image using a Sobel operator. The Canny operator also provides the same quality of result but for this work the Sobel operator provides the better boundary region of a hand gesture with high and low-level intensity values than the other edge-detection operators. Thirdly, feature extraction is used to obtain distinctive features from the given datasets, which reduces the dimensionality of features in different orientations, illumination, and shape using a SURF transform. In this work, the comparison has been made between dimension, recall, and precision for different qualities of feature vector. Fourthly, the hand gesture is constructed using HMM for different classes in the dataset. This could be validated by K-fold cross-validation for simplifying the process of the classification to get the trained and tested data. The model parameter of HMM is then estimated in the defined structure using its feature points. The lower and upper bounds signify the model evidence as to how it can be performed. This produces an average recognition rate of 80.16% and an error rate of 6.79% respectively. In order to avoid time and memory consumption due to the recursiveness in both the forward-backward algorithm and the Viterbi algorithm, the training process was tabulated with a manual relation between the number of states s and the length of the sequence t, which are shown in Table 6.

Once again the same procedure was followed from the validation with its average best cost value of 2,104,646.87 and its CSA initial parameters. The result shows that the average value of performance metrics with CSA is increased by 1.05% of the recognition rate and decreased by 0.62% of the error rate respectively. A comparative analysis was made for the proposed method CSA-HMM with ABC-HMM and GA-HMM. This signifies that the optimized model of the hand gesture recognition system has been improved by the recognition rate and accuracy for virtual reality applications.

References

1. M.A. Al-Betar, β-Hill climbing: an exploratory local search, in *Neural Computing and Applications* (Springer, Berlin, 2016), pp. 1–16
2. H. Ardo, K. Astrom, R. Berthilsson, Real time Viterbi optimization of hidden Markov models for multi target tracking, in *IEEE Workshop on Motion and Video Computing* (2007). http://dx.doi.org/10.1109/WMVC.2007.33
3. H. Bay, T. Tuytelaars, L.V. Gool, SURF: speeded up robust features, in *Proceedings of European Conference ECCV* (2006), pp. 404–417
4. S. Bhowmick, S. Kumar, A. Kumar, Hand gesture recognition of English alphabets using artificial neural network, in *IEEE 2nd International Conference on Recent Trends in Information Systems (ReTIS)* (2015). http://dx.doi.org/10.1109/ReTIS.2015.723913
5. C. Blum, A. Roli, Metaheuristics in combinatorial optimization: overview and conceptual comparison. ACM Comput. Surv. **35**(3), 268–308 (2003)
6. A.L. Bolaji, M.A. Al-Betar, M.A. Awadallah, A.T. Khader, L.M. Abualigah, A comprehensive review: Krill Herd algorithm and its applications. Appl. Soft Comput. **49**, 437–446 (2016)
7. C.T. Brown, L.S. Liebovitch, R. Glendon, Lévy flights in Dobe Ju/'hoansi foraging patterns. Hum. Ecol. **35**(1), 129–138 (2007)
8. T. Cardoso, J. Delgado, J. Barata, Hand gesture recognition towards enhancing accessibility, in *6th International Conference on Software Development and Technologies for Enhancing Accessibility and Fighting Infoexclusion*. Procedia Computer Science, vol. 67 (2015), pp. 419–429
9. Q. Chen, A. El-Sawah, C. Joslin, N.D. Georganas, A dynamic gesture interface for virtual environments based on hidden Markov models, in *IEEE International Workshop on Haptic Audio Visual Environments and Their Applications* (2005). http://dx.doi.org/10.1109/HAVE.2005.1545662

10. J.Y. Cheng, T. Mailund, Ancestral population genomics using coalescence hidden Markov models and heuristic optimisation algorithms. Comput. Biol. Chem. **57**, 80–92 (2015)
11. Y. Dennemont, G. Bouyer, S. Otmane, M. Mallem, A discrete hidden Markov model recognition module for temporal series: application to real-time 3D hand gestures, in *3rd International Conference on Image Processing Theory, Tools, Applications* (2013). http://dx.doi.org/10.1109/IPTA.2012.6469509
12. J. Du, W. Sheng, M. Liu, Human-guided robot 3D mapping using virtual reality technology, in *IEEE/RSJ International Conference on Intelligent Robots and Systems* (2016). http://dx.doi.org/10.1109/IROS.2016.7759680
13. I. Fister Jr., X.S. Yang, D. Fister et al., Cuckoo search: a brief literature review, in *Cuckoo Search and Firefly Algorithm: Theory and Applications* (Springer, Berlin, 2014), pp. 49–62
14. G.D. Forney, The Viterbi algorithm. Proc. IEEE **61**(3), 268–278 (1973). https://doi.org/10.1109/PROC.1973.9030
15. W. Geem, J.H. Kim, G. Loganathan, A new heuristic optimization algorithm: harmony search. Simulation **76**(2), 60–68 (2001)
16. F. Glover, Heuristics for integer programming using surrogate constraints. Decis. Sci. **8**(1), 156–166 (1977)
17. R.C. Gonzales, R.E. Woods, *Digital Image Processing*, 3rd edn. (Prentice Hall, Upper Saddle River, 2008)
18. J.H. Holland, *Adaptation in Natural and Artificial Systems: An Introductory Analysis with Applications to Biology, Control, and Artificial Intelligence* (University of Michigan Press, Michigan, 1975)
19. K. James, E. Rusell, Particle swarm optimization, in *Proceedings of the 1995 IEEE International Conference on Neural Networks* (1995), pp. 1942–1948
20. L. Juan, O. Gwun, A comparison of SIFT, PCA-SIFT and SURF. Int. J. Image Proc. **3**(4), 143–152 (2009)
21. D. Karaboga, An idea based on honey bee swarm for numerical optimization. Technical Report-tr06, Erciyes University, Engineering Faculty, Computer Engineering Department (2005)
22. K. Khan, A. Sahai, Neural-based cuckoo search of employee health and safety (HS). Int. J. Intell. Syst. Appl. **5**(20), 76–83 (2013)
23. S. Kirkpatrick, C.D. Gelatt, M.P. Vecchi et al., Optimization by simulated annealing. Science **220**(4598), 671–680 (1983)
24. J.R. Koza, *Genetic Programming II: Automatic Discovery of Reusable Subprograms* (Cambridge University Press, Cambridge, 1994)
25. S. Koziel, X.S. Yang, *Computational Optimization, Methods and Algorithms*, vol. 356 (Springer, Berlin, 2011)
26. M.A. Laskar, A.J. Das, A.K. Talukdar, K.K. Sarma, Stereo vision-based hand gesture recognition under 3D environment. Procedia Comput. Sci. **58**, 194–201 (2015)
27. R. Li, J.H. Zheng, C.Q. Pei, Text information extraction based on genetic algorithm and hidden Markov model, in *2009 First International Workshop on Education Technology and Computer Science*, vol. 1 (2009), pp. 334–338
28. C.S. Panda, S. Patnaik, Better edge-gap in gray scale image using Gaussian method. Int. J. Comput. Appl. Math. **5**(1), 53–65 (2010)
29. I. Pavlyukevich, Lévy flights, non-local search and simulated annealing. J. Comput. Phys. **226**(2), 129–138 (2007)
30. A. Pradhana, B.B.V.L. Deepak, Design of intangible interface for mouse less computer handling using hand gesture. Procedia Comput. Sci. **79**, 287–292 (2016)
31. F. Qin, A. Auerbach, F. Sachs, A direct optimization approach to hidden Markov modeling for single channel kinetics. Biophys. J. **79**, 1915–1927 (2000)
32. L.R. Rabiner, A tutorial on hidden Markov models and selected applications in speech recognition. Proc. IEEE **77**, 257–285 (1989)
33. R. Rajabioun, Cuckoo optimization algorithm. Appl. Soft Comput. **11**(8), 5508–5518 (2011)

34. J. Rau, J. Gao, Z. Gong, Z. Ziang, Low-cost hand gesture learning and recognition based on hidden Markov model, in *2nd International Symposium on Information Science and Engineering* (2010). http://dx.doi.org/10.1109/ISISE.2009.53

35. K.M. Sagayam, D.J. Hemanth, Hand posture and gesture recognition techniques for virtual reality applications: a survey. Virtual Reality **21**(2), 91–107 (2017)

36. K.M. Sagayam, D.J. Hemanth, Application of pseudo 2-D hidden Markov model for hand gesture recognition, in *Proceedings of the First International Conference on Computational Intelligence and Informatics*. Advanced in Intelligent Systems and Computing, vol. 507 (Springer, Singapore, 2017). https://doi.org/10.1007/978-981-10-2471-9_18

37. K.M. Sagayam, D.J. Hemanth, Comparative analysis of 1-D HMM and 2-D HM for hand motion recognition applications, in *Progress in Intelligent Computing Techniques: Theory, Practice, and Applications*. Advanced in Intelligent Systems and Computing, vol. 518 (Springer, Singapore, 2017). https://doi.org/10.1007/978-981-10-3373-5_22

38. R.P. Sharma, G.K. Verma, Human computer interaction using hand gesture, in *11th International Multi-Conference on Information Processing*. Procedia Computer Science, vol. 54 (2015), pp. 721–727

39. M. Shehab, A.T. Khader, M.A. Al-Betar, A survey on applications and variants of the cuckoo search algorithm. Appl. Soft Comput. (2017). http://dx.doi.org/10.1016/j.asoc.2017.02.034

40. J. Shotton, A. Fitzgibbon, M. Cook et al., Real-time human pose recognition in parts from single depth images. CVPR **2011**, 1297–1304 (2011) https://doi.org/10.1109/CVPR.2011.5995316

41. R. Snapp, CS 295: pattern recognition, course notes. Department of Computer Science, University of Vermont (2015). http://www.cs.uvm.edu/~snapp/teaching/CS295PR/whatispr.html

42. R. Storn, K.V. Price, Minimizing the real functions of the ICEC96 contest by differential evolution, in *International Conference on Evolutionary Computation* (1996), pp. 842–844

43. P. Sykora, P. Kamencay, R. Hudec, Comparison of SIFT and SURF methods for use of hand gesture recognition based on depth map, in *AASRI Conference on Circuits and Signal Processing*, vol. 9 (2014), pp. 19–24

44. R.-L. Vieriu, L. Mironica, B.-T. Goras, Background invariant static hand gesture recognition based on hidden Markov models, in *International Symposium on Signals, Circuits and Systems* (IEEE Publisher, New York, 2013). https://doi.org/10.1109/ISSCS.2013.6651245

45. G.M. Viswanathan, S.V. Buldyrev, S. Havlin et al., Optimizing the success of random searches. Nature **401**(6756), 911–914 (1999)

46. G.M. Viswanathan, F. Bartumeus, S.V. Buldyrev et al., Lévy flight random searches in biological phenomena. Physica A **314**(1), 208–213 (2002)

47. Z. Wang, E. Giannopoulos, M. Slater, A. Peer, Handshake: realistic human-robot interaction in haptic enhanced virtual reality. J. Presence **20**(4), 371–392 (2011)

48. J. Xiao, L. Zou, C. Li, Optimization of hidden Markov model by a genetic algorithm for web information extraction, in *Advances in Intelligent Systems Research. International Conference on Intelligent Systems and Knowledge Engineering* (2007). https://doi.org/10.2991/iske.2007.48,2007

49. X.S. Yang, *Nature-Inspired Heuristic Algorithm* (Luniver Press, Beckington, 2008), pp. 242–246

50. X.S. Yang, *Nature-Inspired Metaheuristic Algorithms*, 2nd edn. (Luniver Press, Frome, 2010)

51. X.S. Yang, A new metaheuristic bat-inspired algorithm, in *Nature Inspired Cooperative Strategies for Optimization* (Springer, Berlin, 2010), pp. 65–74

52. X.S. Yang, Firefly algorithm, in *Engineering Optimization* (2010), pp. 221–230. http://dx.doi.org/10:1002/9780470640425.ch17

53. X.S. Yang, S. Deb, Cuckoo Search via levy flights, in *Nature & Biologically Inspired Computing. IEEE World Congress on NaBIC* (2009), pp. 210–214

54. X.S. Yang, S. Deb, Cuckoo search: recent advances and applications. Neural Comput. Appl. **24**(1), 169–174 (2014)
55. Z. Yang, Y. Li, W. Cheng, Y. Zheng, Dynamic hand gesture recognition using hidden Markov models, in *IEEE 7th International Conference on Computer Science & Education* (2012). http://dx.doi.org/10.1109/ICCSE.2012.6295092

Satellite Image Contrast Enhancement Using Fuzzy Termite Colony Optimization

Biswajit Biswas and Biplab Kanti Sen

Abstract Image enhancement is an essential subdomain of image processing which caters to the enhancement of visual information within an image. Researchers incorporate different bio-inspired methodologies which imitate the behavior of natural species for optimization-based enhancement techniques. Particle Swarm Optimization imitates the behavior of swarms to discover the finest possible solution in the search space. The peculiar nature of ants to accumulate information about the environment by depositing pheromones is adopted by another technique called Ant Colony Optimization. However, termites have both these characteristics common in them. In this work, the authors have proposed a Termite Colony Optimization (TCO) algorithm based on the behavior of termites. Thereafter they use the proposed algorithm and fuzzy entropy for satellite image contrast enhancement. This technique offers better contrast enhancement of images by utilizing a type-2 fuzzy system and TCO. Initially two sub-images from the input image, named lower and upper in the fuzzy domain, are determined by a type-2 fuzzy system. The S-shape membership function is used for fuzzification. Then an objective function such as fuzzy entropy is optimized in terms of TCO and the adaptive parameters are defined which are applied in the proposed enhancement technique. The performance of the proposed method is evaluated and compared with a number of optimization-based enhancement methods using several test images with several statistical metrics. Moreover, the execution time of TCO is evaluated to find its applicability in real time. Better experimental results over the conventional optimization based enhancement techniques demonstrate the superiority of our proposed methodology.

Keywords Image contrast enhancement · Satellite image · Termite colony optimization · Type-2 fuzzy sets

B. Biswas (✉) · B. K. Sen
University of Calcutta, Kolkata, West Bengal, India

© Springer International Publishing AG, part of Springer Nature 2018
S. Bhattacharyya (ed.), *Hybrid Metaheuristics for Image Analysis*,
https://doi.org/10.1007/978-3-319-77625-5_5

1 Introduction

Image enhancement is one of the crucial parts in the domain of image processing. The objective of image processing is not confined to the improvement of
the visual quality of an image but also deals with the enhancement of visual
information within an image for better human perception [15, 19, 22, 25, 28, 36].
An efficient image enhancement technique plays an immensely important role in
the field of computer vision, biomedical image analysis, surveillance applications,
robot navigation, weapon detection, and so on [15, 19, 22, 36]. The common
prevalent image enhancement methodologies of recent times are histogram and
transform-based techniques [1, 3, 19, 31, 34], contrast stretching [19, 22, 36], image
smoothing [19, 25, 32], image sharpening [7, 22], inverse filtering [25, 32, 36], and
Wiener filtering [7, 28]. These are widely used but lack success when implemented
in the domain of colored images [15, 19, 22, 28]. Another popular enhancement
technique is Hue preserved color image enhancement, which extends existing gray
scale contrast intensification procedures in color images [10, 15, 19, 22].

Satellite images play a significant role in numerous areas at the present time [11,
12, 19, 34, 36]. Due to the growing application to high-resolution images in
satellite image processing, contrast enhancement techniques are widely used to
achieve better visual perception and color reproduction [11, 12, 15, 36]. Generally,
unprocessed satellite images (i.e. initially captured raw data) have covered a
comparatively narrow range of brightness values; in this case, contrast enhancement
is commonly used to improve a narrow range of multi-band satellite images for
better understanding and imaging [11, 12, 19, 34, 36]. On the other hand, the
average resolution of raw satellite images is very low due to various factors such as
absorption, scattering, illumination of the light source, and the lens apparatus. For
several decades, various enhancement techniques have been suggested for increased
resolution and contrast of the satellite images. On the other hand, waveletbased
techniques have been established as the most robust process that satisfies the
required purpose. In the literature, different histogram equalization techniques have
been proposed for image enhancement. However, histogram equalization is the
most common approach for enhancing the contrast in the area of different imaging
applications, for example, surveillance, medical image processing, and satellite
image processing [11, 12, 32, 34]. However, they often reduce the overall image
quality (color information, illumination, etc.) by showing certain artifacts in both
low and high-intensity regions in the images.

For low contrast image processing, several image enhancement techniques have
been utilized to overcome the leading problem of image enhancement, such as
improved image contrast, brightness, and the preservation of color information. In
this regard, a collection of different image enhancement techniques with different
ideas has successfully been used in satellite imaging in the literature [1, 9, 11,
12, 25, 28]. Similarly, there are several histogram based techniques that have
been suggested as able to overcome these problems [11, 12, 19, 34, 36], such
as generalized histogram equalization (GHE) [12, 19, 35] and local histogram
equalization(LHE) [20, 32, 34] and etc. [9, 17, 21, 26]. The GHE technique is a

simple and effective approach for contrast improvement for many image processing applications. LHE is better than GHE, but suffers from inappropriate intensity distribution of the pixel in the image areas. However, most of them cannot maintain an average brightness level, which creates artifacts (either under or over-saturation) in the resulting image [11, 12, 15]. Most commonly satellite images are certainly affected by the most prevalent contrast enhancement techniques, for example, drifting pixel intensity, saturation of brightness levels, and distorted color details. The number of artifacts which are not propagated throughout the image in the sense of both spatial locations and intensity levels needs to be minimized [11, 12, 15, 22]. In most of the cases, enhancement algorithms for satellites reduce the pixel distortion in low and high-intensity regions. The proposed algorithm solves this problem by using the advanced fuzzy-based metaheuristic approach [8, 9, 17, 21, 22, 36].

In recent years, many efficient methods have been developed for image enhancement, such as the bat algorithm [5, 17, 19], the cuckoo algorithm [5, 8, 14], and the immune algorithm [2, 19, 33]. In [15, 17, 19], the authors combined the neuron network and the bat algorithm to explain the image enhancement problem, and applied the bat algorithm to tune the parameters of the modified neuron model for the maximization of the indices contrast enhancement factor and mean opinion score. In [5, 8, 14], the author suggested image enhancement by the cuckoo algorithm and optimum wavelet by choosing the best wavelet coefficients. Recently, many researchers have suggested excellent image enhancement methods [15, 16, 19, 36]. Interestingly, we found that most suggested methods include Principal Component Analysis [3, 16, 19, 23], Independent Component Analysis [1, 15, 16, 32], Sparse Coding [5, 19, 23], and Gaussian Mixture Model [7, 19, 20, 25]. Most recently, different deep learning based schemes, such as Stacked Denoising Sparse Auto-encoder (SSDA), have been applied to image enhancement. However, the author has been used the SSDA in image enhancement but for large number of training data, the model suffered over-fitting and the trained model parameters restricted under input dataset directly. In [9, 13, 15, 20], the authors have been suggested and overcame some limitations of the SSDA. The limitations are removed by using the end-to-end processing approach and the semantics of feature extraction are performed on the basis of predefined parameters.

Various neural network-based image enhanced schemes are straightforwardly recommended for image processing in the literature [13, 18, 20, 22, 24, 36]. Recently, deep convolutional networks (DCNN) have realized significant growth on low-level vision and image processing tasks, such as depth estimation [13, 18, 22, 24, 36], optical flow [18, 36], super-resolution [13, 18], demosaicking and denoising [13, 18, 24], image matting [13, 24], colorization [22, 24, 36], and general image-to-image translation tasks [18, 36]. The latest approaches have even explored learning deep networks contained by a bilateral grid [13, 18, 24]; however, this method does not achieve the proper task of learning image transformations in that learning space, but rather emphasizing classification and semantic segmentation. Typically, the DCNN architectures have been trained more or less to approximate a general class of supervises in the gradient domain under a multi-layer network [13, 18]. In [13, 22], the author designed a DCNN to train recursively filters for color interpolation, denoising, image-smoothing, and painting [13, 20]. However, they mutually train a group of recursive networks and a convolutional network to calculate image-dependent promulgation weights [13, 18]. On the other hand, we can notice that some of the DCNN works for the low resolution images on GPU environment but *still they are excessively slow for the real-time processing of high-resolution image applications.*

Further, several efficient fuzzy enhancement methods have been developed for low light images on the basis of the statistical features of the gray-level histogram [8, 16, 19, 22, 27, 34]. Fuzzy intensification methodologies are used to enhance images by the authors in [9, 15, 17, 21]. The authors have been suggest a fuzzy system to enhance an image by using the Hue-Saturation-Value (HSV) model and considered the Gaussian parameters as Saturation (S) and Value (V), respectively in [15, 17]. In this work, the S-curve is the transformation function; by optimizing the image information through entropy [9, 15, 17, 21] the parameters of the S-curve are estimated. However, the problem lies in their dependency on several image properties like local surface reflectance, occlusion, and shadow to enhance the quality of the image [3, 16, 19, 22]. Authors have proposed a fuzzy method to enhance dark regions and that also works well with night images [15, 16]. In [15, 19, 22] the authors used fuzzy membership functions for the enhancement of color images. Another generalized iterative fuzzy enhancement algorithm is proposed in [9, 15, 17, 21] for the degraded images with less gray levels and low contrasts. Here the statistical features of the gray-level histogram of images are used as an image quality assessment criterion to control the iterative procedure. In [12, 31, 33], the author used a fuzzy system to enhance an HSV image by modeling S and V as Gaussian. In [9, 15, 17, 19, 21], the authors have successfully explored the relationship between intensity and saturation. This work also proposed a new method called SI correction for removing color distortion in the HSV space. Some work [9, 15, 17] used a scene irradiance map as an image property to enhance the image with many factors, such as local surface reflectance, occlusion, and shadow.

Recent research has exposed different measures of image quality and the widespread use of gray-level image enhancement techniques [9, 15, 17, 21]. The most popular measures among these constitute the activity measurement of edge pixels, pixel intensity variation, and variation of image entropy [2, 19, 22, 34]. Such measures have been successfully utilized in the domain of image enhancement incorporated with swarm intelligence optimization techniques like the Genetic algorithm [4, 19, 22, 23, 30], the Differential Evaluation Optimization algorithm [4, 8, 22, 30], Particle Swarm Optimization (PSO) [2, 22, 29, 37], the Cuckoo Search (CS) algorithm [6, 10, 19], Artificial Bee Colony (ABC) [5, 8, 14], Ant Colony Optimization (ACO) [15, 17, 19, 22, 33] and Bacterial Foraging (BF) [19, 22, 33]. In the case of the BF algorithm, this has been used by the authors in [30, 33] to optimize the objective function for color image enhancement. ABC is a metaheuristic technique which is efficiently utilized for the production of high-quality image enhancement methodologies. This is a comparatively new metaheuristic approach which has proved to be efficient over the existing techniques in solving optimization algorithms. This is shown by the authors in [15, 17, 19, 22, 33].

Resolving enhancement problems by the Artificial Ant Colony System or the ACO technique has been successfully illustrated by the authors in [15, 17, 19, 22, 33]. The authors have been successfully introduced an image contrast enhancement technique with the combination of particle swarm optimization (PCO) and cuckoo search (CS) algorithms to solve the high-quality image enhancement problems [2, 22, 29, 37]. Like the Bat optimization algorithm [6, 22, 33], the Genetic Algorithm (GA) [4, 19, 22, 23, 30], and the Differential evaluation algorithm (DE) [4, 8, 22, 30], bio-inspired intelligent based techniques also find an important place in the domain of image enhancement [15, 19, 22, 32]. In [9, 15, 17, 23, 33], the authors have proposed a GA technique that utilizes the algorithm to find an optimal mapping onto the gray levels of the source image. This results in the production of an enhanced image with new gray levels and better contrast image quality. The authors in [4, 6, 30] proposed another GA based color image enhancement approach which was used in nonlinear transforms as a function for contrast enhancement. In [6, 15, 17, 19, 23], the author suggested a DE algorithm to solve image enhancement problems. The DE algorithm is more local optimal than GA but parametric restriction is more sensitive than DE under maximal optimization [4, 19, 23, 30]. In this work, we have analyzed the behavior of termites and expose the common nature which they share with ants and swarms. We propose a new technique for image enhancement, Termite Colony Optimization (TCO), which is based on the nature and action of termites in the real world. TCO is a combination of PSO and ACO techniques which when utilized for image enhancement provides better results than prevalent techniques. The proposed method (TCO) is used to optimize fuzzy entropy in a fuzzy domain. Entropy used here is an objective function which is used to find the optimized value of the parameters in the domain of image enhancement. The key contributions of this study are briefly:

1. A bio-inspired panchromatic (PAN) satellite image enhancement scheme is proposed and studied.
2. The proposed model effectively uses the standard S-fuzzy membership function and Shanon fuzzy entropy measure to optimize the image contrast within the dynamic range of gray levels.
3. A special mechanism of the proposed TCO is employed to optimize the quality of sharpness in the low contrast image regions in terms of the function optimization criterion.
4. Simulation of the proposed technique and outstanding performance is offered.
5. A collection of PAN satellite image data examples illustrate the practical usefulness of the proposed model.

The work is organized as follows. We introduce and explain the basic concepts of TCO in Sect. 2. In Sect. 3, we discuss the methodology of TCO application in fuzzy image enhancement. Implementation and results are discussed in Sect. 4 with conclusions in Sect. 5.

2   Termite Colony Optimization Algorithm

ACO [22, 23, 36] and PCO are population-based optimization algorithms [22, 36, 37]. The characteristics of termites have a similarity with the behavior of ants, depicted by ACO, and the nature of swarms is represented by PCO [24, 25]. TCO is an evolutionary computing algorithm inspired by swarm and ant intelligence. It resembles the behavior of swarms to find the best possible position in the search space for an optimal problem. Each termite moves with a velocity following the direction of its previous best position. Hence, like the swarm, it moves towards the best previous position in the search space. Moreover, termites are also guided by pheromones as are ants. The pheromone trail leads them to accumulate information about the environment. This trail can be accessed by other members of the colony, like ants.

For an image enhancement problem, we assume the initial position of a termite to be i in a termite colony. The probability of the termite selecting a route from pixel i to pixel j among n pixels is given by [6, 14, 33, 37]

$$\tau_{ij} = \frac{\tau_{ij}^{\alpha} d_{ij}^{\beta}}{\sum_{i,j=1}^{n} \tau_{ij}^{\alpha} d_{ij}^{\beta}} \tag{1}$$

For each termite, the parameter α controls the relative evaporation or delay of pheromone intensity and the parameter β controls the desirability decision. Both α and β is > 0. τ_{ij} represents the concentration of pheromone associated with edge (i, j), and d_{ij} is the desirability of edge (i, j). Due to evaporation, the concentration of pheromone decays with time. For simplicity, the pheromone is updated as follows [22, 33, 36]:

$$\tau_{ij}(t+1) = (1-\gamma)\tau_{ij}(t) + \delta\tau_{ij}(t) \tag{2}$$

where $\gamma \in [0, 1]$ is the rate of decay of the pheromone intensity. The increment $\delta\tau_{ij}(t)$ is the amount of pheromone deposited at time t along route i to j when a termite travels along a path. The pheromone update policy and the decision rule exploited within the termite system are given by Eqs. (2) and (1), respectively. Each termite deposits a pheromone at all chosen edges and thus the additional pheromone deposited at each edge (i, j) is given by

$$\delta\tau_{ij}(t) = \sum_{k=1}^{m} \delta\tau_{ij}^{k}(t) \tag{3}$$

where m is the number of termites and $\delta\tau_{ij}^{k}(t)$ is the additional pheromone deposited at edge (i, j) by the kth termite at the end of iteration t.

Assuming that the size of the swarm is N within the M dimensional search space, then the position of the ith termite is denoted as $X_i\left(x_{i_1}, x_{i_2}, \ldots, x_{i_M}\right)$, which indicates a possible solution of an optimal problem. The velocity of each termite is represented by $V_i \left(v_{i_1}, v_{i_2}, \ldots, v_{i_M}\right)$. The best previous position of the termite is denoted as $P_i \left(p_{i_1}, p_{i_2}, \ldots, p_{i_M}\right)$, while the best previous position of the whole swarm is represented as $P_g \left(p_{g_1}, p_{g_2}, \ldots, p_{g_M}\right)$. The termites exhibit characteristics denoted by the following equations [17, 33]:

$$v_{im}^{k+1} = \omega^k * v_{im}^k + c_1 * \tau_{ij}^k (t) * \left(p_{im} - x_{im}^k\right) \Delta t + c_2 * \tau_{ij}^k (t) * \left(p_{gm} - x_{im}^k\right) / \Delta t \quad (4)$$

$$x_{im}^{k+1} = x_{im}^k + \Delta t * v_{im}^k \quad (5)$$

$$\omega^k = \omega_{\max} - k * (\omega_{\max} - \omega_{\min}) / k_{\max} \quad (6)$$

where $1 \leq m \leq M$, and $rand()$ are dimension of the termite and random number with uniform distribution $U(0, 1)$; c_1 and c_2 are acceleration coefficients; ω is the inertia weight; $\omega_{\max}$ and $\omega_{\min}$ are the maximum and minimum values of ω, respectively; k and $k_{\max}$ are the current and the maximum iteration times respectively; generally Δt is the time unit. v_{im}^{k+1} and x_{im}^{k+1} must be under the restricted conditions as follows [6, 33, 36, 37]:

$$v_{im}^{k+1} = \begin{cases} v_{im}^{k+1} & -v_{\max} \leq v_{im}^{k+1} \leq v_{\max} \\ v_{\max} & v_{im}^{k+1} > v_{\max} \\ -v_{\max} & v_{im}^{k+1} < -v_{\max} \end{cases} \quad (7)$$

$$x_{im}^{k+1} = \begin{cases} x_{im}^{k+1} & x_{\min} \leq x_{im}^{k+1} \leq x_{\max} \\ x_{init} & x_{im}^{k+1} > x_{\max} \\ x_{init} & x_{im}^{k+1} < x_{\min} \end{cases} \quad (8)$$

$$x_{init}^{k+1} = x_{\min} + \gamma * (x_{\max} - x_{\min}) \quad (9)$$

where $v_{\max}$ is the maximum value of v; $x_{\max}$ and $x_{\min}$ are the maximum and minimum values of x, respectively (Fig. 1).

The basic steps of TCO can be summarized as follows:

1. Initialize the population with weights, position of swarm termites, and the number of training iterations. Every swarm termite has its distinct random position, velocity, desirability, and rate of evaporation of the pheromone.
2. Verify the fitness function value for each termite swarm.
3. Determine the best position and evaporation rate of the pheromone of each termite swarm.
4. Determine the position of the best termite swarm.
5. Update the evaporation rate of the pheromone, velocity, and position of each termite swarm by Eqs. (2), (7), and (8) respectively.
6. Stop if the condition of optimization is satisfied. If not, repeat from Step 2.

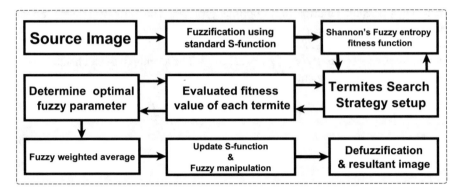

Fig. 1 A schematic overview of the proposed algorithm. All steps correspond to Algorithm 1

In our previous explanation of our algorithm TCO, we discussed the convergent properties of the TCO in brief. The TCO algorithm may have a formal proof of convergence like other metaheuristic approaches (such as ACO and swarm optimization [9, 17, 21, 27]) which depends on the size of the termite. We can control the behavior of the TCO algorithm by manipulating the termites within the arbitrary position. Any given termites with size N and position P within a sufficient number of iterations always lead to equilibrium; and at that point the pheromone distribution of the accepted states is stationary. We may note that for size N and position P almost any change is accepted. This equilibrium convenience state indicates that the TCO algorithm is able to expand a large neighborhood of the current position of the termites. Initially, position P transitions to the next position with the same termite size N, which becomes less frequent, and the solution stabilizes at the stationary states. The TCO algorithm depicts different advantages over the other metaheuristic approach, for example,

1. For larger size problems, the transitions overhead are significantly reduced.
2. The position of transitions depends not only on the size of the termites but also the position of its states. Thus, in comparison to other metaheuristic approaches, as the size of the problem increases, the performance of the method improves.
3. For optimal and stationary states, fewer moves are required for termites to be transferred because the TCO algorithm converges and reaches an equilibrium state at a certain number of iterations, for the reason that the efficiency of the TCO algorithm increases.

Figure 2a shows the convergent graph of the TCO algorithm with a certain number of iteration steps. On the basis of the following explanation, we apply the TCO algorithm in low contrast satellite PAN image contrast enhancement: (1) we address the scheme of the transformation function in the fuzzy domain, which creates new fuzzy pixel intensities for the improved contrast image from the original source image; and (2) we define the proper fitness function which investigates the quality of the produced image. In this study, we choose fuzzy entropy as the fitness function. These two challenges are described in the following sections.

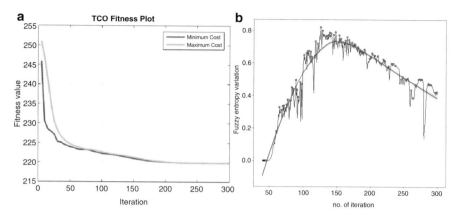

Fig. 2 Illustration of convergence and variation of TCO with a given number of iterations: (**a**) Plot for TCO convergence; (**b**) Plot for Fuzzy entropy variation

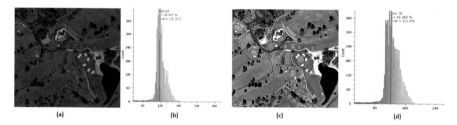

Fig. 3 Comparative histograms of a given PAN image and the result of the TCO. (**a**) Original image, (**b**) histogram of the original image, (**c**) enhanced by the TCO, (**d**) histogram of the output image

3   Fuzzy TCO Application in Image Enhancement

Occasionally, images of the scenes appear in nature contrasting with their visual comprehension, which is why the histogram of an image is unable to occupy the entire dynamic range. It is well known that the image region turns out darker when the intensity distribution is skewed towards the lower part of the histogram as well as the image appearing brighter [17, 22, 36] when it is skewed towards the upper part of the histogram and the image seems to be perceived as blurred. This means that those images have regions which are over dark and over bright, which indicates that a group of neighborhood pixels has gray levels very close to either the minimum or the maximum with respect to the available dynamic range. In this regard, we utilized the available dynamic range of a given satellite image by means of a fuzzy based metaheuristic enhancement technique (TCO). In Figure 3 (a–d), we have illustrated the original and extended histograms of the source and enhanced images of a ground truth PAN satellite image.

GHE based contrast enhancement techniques are not able to conserve boundary or edge details and cannot conceal certain artifacts in the low and high-intensity regions, even though GHE is considered one of the best enhancement methods. However, sometimes it shows definite artifacts, for instance, false color composites, less brightness, darkness, and irregular dim regions according to the nature of the images. If spatially varying intensity distributions are not considered then the corresponding enhanced images may have intensity distortion and may lose image details in some regions.

In this study, we have proposed suitable fuzzy image enhancement techniques which deal with the fuzziness and uncertainty of images [16, 17, 27, 36]. Let X refer to an image of size $M \times N$ with L dynamic gray levels ranging from L_{min} to L_{max}. x_{ij} represents the gray level of the (i, j)th pixel in X [22]. Depending on fuzzy set theory, X is transformed into a set of fuzzy singletons G using a particular membership function.

$$G = \{\mu_X(x_{ij})\ i = 1, 2, \ldots, M;\ \ j = 1, 2, \ldots, N\} \tag{10}$$

where $0 \leq \mu_X(x_{ij}) \leq 1$ and $\mu_X(x_{ij})$ represent some image properties like brightness and grayness for the (i, j)th pixel.

Fuzzification is a transformation of the values of intensity to an interval between 0 and 1 [17, 22, 27, 36]. This is done by the use of any suitable fuzzy membership function. The standard S-function is commonly used as a membership function due to its simplicity and robustness. This is defined as [17, 22, 27, 36]:

$$S = \begin{cases} 0 & x < p \\ 2 \times \left(\frac{x-p}{r-p}\right)^2 & p \leq x < q \\ 1 - 2 \times \left(\frac{r-x}{r-p}\right)^2 & q \leq x < r \\ 1 & x \geq r \end{cases} \tag{11}$$

where p, q and r denote fuzzy parameters.

The parameter q is the cross-over point and is given by $q = (p + r)/2$. The interval $[p, r]$ is the fuzzy region whose width is $2\Delta q = r - p$. The interval $[L_{min}, p]$ and $[r, L_{max}]$ are defined as non-fuzzy regions.

Conventionally, Type-2 fuzzy sets deal with the uncertain assignment of a membership degree [17, 22, 27, 36]. So, it is more consistent than a Type-1 fuzzy set. In general, the amount of uncertainty in the membership of a Type-2 fuzzy set is represented by the footprint of uncertainty, which is described in terms of upper and lower membership functions. According to a Type-2 fuzzy set, we can define a Type-1 fuzzy set and assign upper and lower membership degrees to each element to determine the footprint of uncertainty [17, 22, 27, 36]. Conventionally, a Type-2 fuzzy set is represented as:

$$\tilde{G} = \{(x, \mu_L(x), \mu_U(x)|),\ \ \forall x \in X,$$
$$\mu_L(x) \leq \mu(x) \leq \mu_U(x), \mu \in [0, 1]\} \tag{12}$$

where $\mu_L(x)$ and $\mu_U(x)$ are the upper and lower membership degrees of the primary membership function $\mu(x)$. Hence, the lower and upper membership values can be defined as [22, 27, 36]:

$$\mu_L(x) = \left[\mu(x)^{q/2}\right], \text{ and } \mu_U(x) = \left[\mu(x)^{2*q}\right] \tag{13}$$

where $r \in [0, 1]$. Based on the lower and upper membership degrees $\mu_L(x)$ and $\mu_U(x)$ of the primary membership function $\mu(x)$, the decomposed membership degrees are composed of the fuzzy weighted average which represents the output image μ_{g_e}. The output image μ_{g_e} is expressed as

$$\mu_{g_e}(x) = (\eta \times \mu_L(x) + (1 - \eta) \times \mu_U(x)) \tag{14}$$

where parameter η is defined as $\eta = 4 * q^2$. It is observed that the brightness of the image depends on the value of η.

Based on the optimal problem, we have constructed the termites in the swarm with position and velocity, which are both two-dimensional vectors of real numbers. There are three main procedures in the proposed image enhancement method: (1) devise the termites in the swarm, (2) design the fitness function, and (3) decide on the swarm search strategy [3, 10, 27]. The detailed procedure is:

3.1 Devise the Termites in the Swarm

The position of each termite $X_i \left(x_{i_1}, x_{i_2}, \ldots, x_{i_M}\right)$ denotes a possible solution of the optimal problem, where M is equal to $M \times N$. However, the velocity of each termite belongs to the same direction with a different velocity. The different position and velocity of each termite is a real number. In this work, we have introduced TCO into the process of primary fuzzy set $\mu_X(x)$ to find the best parameter adaptively. The parameters of the proposed method are the size of the swarm, which is $M \times N$, and the maximum iterative time is 200, $c_1 = 1.1915, c_2 = 1.4431, x_{max} = \max(\mu_X(x))$, $x_{min} = \min(\mu_X(x))$. The mutation rate is 0.175 and the range of the local search is ± 30 around P_g and $\alpha = 0.25, \beta = 0.5, \gamma = 0.4551$ respectively.

Now, we consider the position of all termites in the swarm colony by a position matrix X_s and the velocity of all termites by a velocity matrix V_s. Concentration of the pheromone is represented by P_s and desirability by D_s for all termites. Now, X_s, V_s, P_s, and D_s are defined as:

$$X_s = \begin{pmatrix} x_{11} & x_{12} & \cdots & x_{1M} \\ x_{21} & x_{22} & \cdots & x_{2M} \\ \vdots & \vdots & \vdots & \vdots \\ x_{N1} & x_{N2} & \cdots & x_{NM} \end{pmatrix} \tag{15}$$

$$V_s = \begin{pmatrix} v_{11} & v_{12} & \cdots & v_{1M} \\ v_{21} & v_{22} & \cdots & v_{2M} \\ \vdots & \vdots & \vdots & \vdots \\ v_{N1} & v_{N2} & \cdots & v_{NM} \end{pmatrix} \tag{16}$$

$$P_s = \frac{1}{X_s}, \text{ and } D_s = \left(Xs_{i+1,j+1}{}^2 + Xs_{i,j}{}^2 \right)^{1/2} \tag{17}$$

where N is the number of termites in the swarm and M is the dimension of each termite. Finally, the enhanced image $G_e(i, j)$ is obtained by the following defuzzification technique [17, 22, 27, 36]:

$$G_e(i, j) = \bigcup_{i=1}^{M} \bigcup_{j=1}^{N} \mu_{g_e}(i, j) * (L - 1) \tag{18}$$

where $G_e(i, j)$ represents the gray level of the (i, j)th pixel in the enhanced image and L denotes the highest gray level of image X.

3.2 Fitness Function Designing

The position of a termite is measured by a special function which is known as a fitness function [17, 21, 27]. The larger the value of the fitness function of a termite, the better the optimal solution becomes. Entropy can be used as a qualitative measure of image quality. The fuzzy contrast of an image is a measure of the total amount of variation of the membership values of intensities of pixels of an image from the cross-over point. Fuzzy entropy is a measure of the uncertainty of a fuzzy set. The fitness function for the optimal solution can be defined as:

$$H(X) = \frac{1}{MN \ln 2} \sum_{i=1}^{M} \sum_{j=1}^{N} S_n \left(\mu_X \left(x_{ij} \right) \right) \tag{19}$$

where $S_n(\cdot)$ is Shannon's function defined by:

$$S_n \left(\mu_X \left(x_{ij} \right) \right) = -\mu_X \left(x_{ij} \right) \ln \mu_X \left(x_{ij} \right) - \left(1 - \mu_X \left(x_{ij} \right) \right) \ln \left(1 - \mu_X \left(x_{ij} \right) \right) \tag{20}$$

Based on the fitness function, the larger functional value represents the more informational quantity of $\mu_X(x)$. Therefore, when the fitness function has the maximum value, the corresponding parameter is the best one. Figure 2b shows the fuzzy entropy optimization graph of the TCO algorithm with a given number of iteration steps.

3.3 Termites Search Strategy

For each image $\mu_x(i, j)$, the swarm search strategy is illustrated as follows:

1. Initialize the position matrix X and the velocity matrix V of the termite swarm by using Eq. (15), where the elements in the position matrix X and the velocity matrix V are initialized according to the following equations:

$$x_X = x_{min} + (x_{max} - x_{min}) * \gamma \tag{21}$$

$$v_X = -v_{max} + 1.25 * v_{max} \tag{22}$$

where $i = 1, 2, \ldots, N$ and $m = 1, 2$; N is the size of the swarm; γ is the real number with the value of 0 or 1; v_{max} is the maximum value of v; x_{max} and x_{min} are the maximum and minimum values of x, respectively; $x_{max} = L_{max}^X$, $x_{min} = L_{min} + 1$; L_{max} and L_{min} are the maximum and minimum values of X respectively.

2. Estimate the fitness value of each termite in the swarm using the fitness function Eq. (19).
3. Update the position of each termite in optimal space. Compare the calculated fitness value of each termite with the fitness value of its best previous position. If the current value is better, then set the current position as its best previous position.
4. Compare the evaluated fitness value of each termite with the fitness value of the whole swarm's best previous position P_g (Eq. (4)). If the current value is better, then set the current position as the whole swarm's best previous position.
5. Calculate the position and velocity of each termite according to Eqs. (21) and (22).
6. The position of each termite is represented by the primary fuzzy membership grades as $\mu_X(x)$. However, to maintain the original proportion in the swarm according to the mutation rate, apply the mutation function to them as follows:

$$mut(x_X) = x_X \times \left(1 + \frac{1}{1 + \exp{-(\sigma)}}\right) \tag{23}$$

where $mut(x_X)$ is the position after mutation and σ denotes the variation of x_X. If the fitness value of the termite after mutation is better than the fitness value of the termite before mutation, the mutation result is updated. Otherwise, the mutation result is rejected.

7. Exit the loop if the stop criterion is satisfied under the predefined maximum iteration time and search if the output of the best parameter P_g is optimal. Otherwise, go to Step (2).
8. Determine the best fuzzy parameter combination (p, r) in the fuzzy and non-fuzzy regions around P_g.
9. Compute the fuzzy parameter r using the best fuzzy parameter combination (p, r) and enhance the image using Eq. (14).

Algorithm 1: Pseudo-code of the proposed algorithm

Require: Image matrix $I_{M \times N}$, $(p_1, p_2, \ldots, p_n) \in I_{M \times N}$;
Initialize parameters c_1, c_2, x_{max}, x_{min}, α, β, γ, τ and search rang set as -30 to $+30$;
set maximum number of iterations n;
Output: Enhanced satellite image I_e;
Preparation: Fuzzify the input matrix $I_{M \times N}$ into $I_{\mu_{ij}}$ fuzzy matrix using $S(p, q, r)$−function
(Eq. (11)) with arbitrary values (e.g. $p = 50$, $q = 75$, $r = 100$);
Initialize matrix X_s, V_c, P_s, D_s by TCO and fuzzy entropy $H(\cdot)$ (Eq. (19));

    while $(k \leq n)$ **do**

        Calculate x_X^k and v_X^k where x_X, v_X are position and velocity matrices according to
        Eqs. (21)–(22);
        Compute fuzzy fitness value $H^k(\cdot)$ by using Eq. (19) with Eq. (4);
        Estimate x_X^{k+1}, v_X^{k+1} for $I_{\mu_{ij}}^k$ according to TCO;

        Update $P_s^k = \frac{1}{X_s^k}$ by Eq. (17);

        $x_X^{k+1} = x_{min}^k + \left(x_{max}^k - x_{min}^k\right) * \gamma$ by Eq. (21);
        $v_X^{k+1} = -v_{max}^k + 1.25 * v_{max}^k$ by Eq. (21);
        Search fuzzy $S(p, q, r)$ parameter (p, q), (q, r) using x_X^{k+1}, v_X^{k+1} when P_s^{k+1} maximum;
        Update $I_{\mu_{ij}}^k$ with parameter (p, q), (q, r) using Eq. (11);
        $k = k + 1$;

    end while

Compute fuzzy weighted average μ_{g_e} by Eq. (14);
Perform fuzzy membership manipulation on μ_{g_e} according to Eq. (13);
Post-process (defuzzification of μ_{g_e}) results and visualization (enhanced image I_e);

---

4 Experiment and Results Analysis

The proposed methodology was implemented on *Intel(R) Core (TM) i5 − 3304S CPU at* 2.80 GHz *with* 16 GB *RAM using Python version anaconda* 4.6. The proposed algorithm was validated by applying it to a set of test images (http://earthobservatory.nasa.gov/). Experiments were performed on over 100 selected Panchromatic QuickBird (http://earthobservatory.nasa.gov/) (*PAN-QuickBird,* 0.7 − *m resolution,* 512 × 512 pixels after resizing) images from various sources which confirm the results. We compared our proposed technique with four metaheuristic methods such as *GA* [4, 23, 30], *DE* [4, 8, 30], *ABC* [5, 8, 14], *PSO* [2, 29, 37], *ACO* [4, 15, 33, 36], and *CSO*[6, 10]. Initially, to enhance the original PAN satellite image, for all aforementioned algorithms have been tested on the three color channels i.e. read (R), green (G) and blue(B) of a color image (RGB) separately. Each channel is considered a 2D gray-level image (monochromatic). Then, results achieved from each enhanced channel were combined to form the resultant color image.

In this study, the above mentioned metaheuristic based approaches are performed as enhancement tasks with adjustments to their own control parameters because these restrict the systematic model of each algorithm. For example, in this work, we used the essential control parameters for the GA and DE approaches as suggested in [4, 23, 30] and [4, 8, 30] using the author's setting. The optimality of the GA algorithm is influenced by different control parameters such as cross-over probabilities and mutation rates. On the other hand, DE is controlled by certain operators such as the initial population generation, cross-over, mutation, and selection. The main steps of all algorithms revolve around the initial population generation, evaluation of the individuals in the population, evaluation of the fitness function, and global optimization. The operation of each process in the TCO is repeated until a predefined stopping criterion such as a maximum generation number that is satisfied for each generation. All the above algorithms are straightforwardly implemented in Python with entropy as the global optimization function and $n = 50$ fixed iterations.

From the above discussion, we can notice that the aforesaid metaheuristic based image enhancement approaches are required to adjust their associated parameters. Due to their parametric based optimization nature, each algorithm can be impacted in their result by their corresponding parameter's values. On the other hand, the parameters of the PSO method are set as follows: number of particles = 30, number of iterations = 50, mutation probability value $p_a = 0.25$, scale factor $\beta = 1.0$ as suggested in [2, 29, 36, 37]. Similarly, the parameters for ABC are set to the following values: total number of bees = 30 and the number of iterations = 50, where employed foragers and idle foragers share half the population respectively as suggested in [5, 8, 14, 22, 36]. For ACO [22, 33, 36] and CSO [6, 10, 36], we set the parameters as follows: number of fireflies = 30, number of iterations = 50, attractiveness probability value $\beta_0 = 0.25$, cooling factor $\kappa = 0.40$, and $\alpha = 0.05$, $\gamma = 1.0$, and $\delta = 0.01$, respectively as suggested in [6, 10, 22, 33, 36].

This section we have described exactly how to use all algorithms to solve low contrast image enhancement, for instance, in Figs. 4(a), 5(a), 6(a), 7(a), 8(a), and

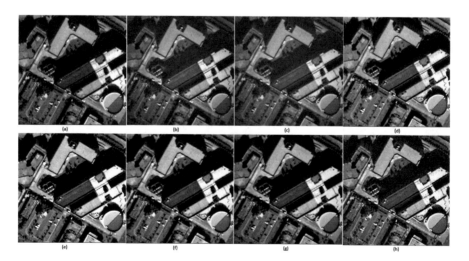

Fig. 4 Simulation results of the source PAN QuickBird image (QuickBird, 0.7-m resolution, 512× 512 pixels after resizing) and enhanced results of different methods. (**a**) Original "PAN image", (**b**) GA-based technique, (**c**) DE-based technique, (**d**) ABC technique, (**e**) PSO-based technique, (**f**) ACO-based technique, (**g**) CSO-based technique, and (**h**) the proposed technique (TCO)

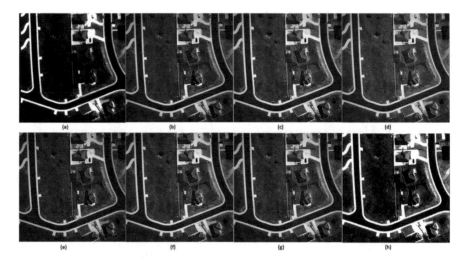

Fig. 5 Simulation results of the source PAN QuickBird image (QuickBird, 0.7-m resolution, 512×
512 pixels after resizing) and enhanced results of different methods. (**a**) Original "PAN image",
(**b**) GA-based technique, (**c**) DE-based technique, (**d**) ABC technique, (**e**) PSO-based technique,
(**f**) ACO-based technique, (**g**) CSO-based technique, and (**h**) the proposed technique (TCO)

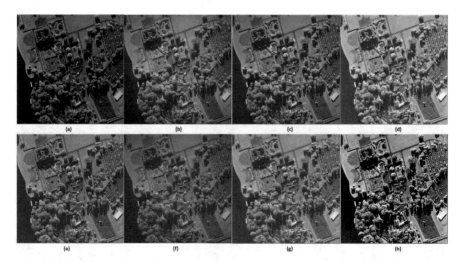

Fig. 6 Simulation results of the source PAN QuickBird image (QuickBird, 0.7-m resolution, 512×
512 pixels after resizing) and enhanced results of different methods. (**a**) Original "PAN image",
(**b**) GA-based technique, (**c**) DE-based technique, (**d**) ABC technique, (**e**) PSO-based technique,
(**f**) ACO-based technique, (**g**) CSO-based technique, and (**h**) the proposed technique (TCO)

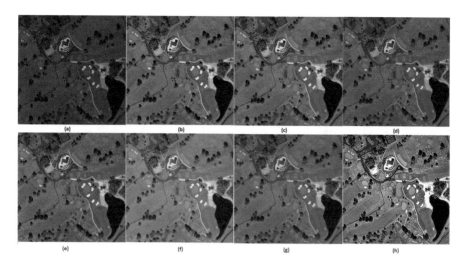

Fig. 7 Simulation results of the source PAN QuickBird image (QuickBird, 0.7-m resolution, 512×512 pixels after resizing) and enhanced results of different methods. (**a**) Original "PAN image", (**b**) GA-based technique, (**c**) DE-based technique, (**d**) ABC technique, (**e**) PSO-based technique, (**f**) ACO-based technique, (**g**) CSO-based technique, and (**h**) the proposed technique (TCO)

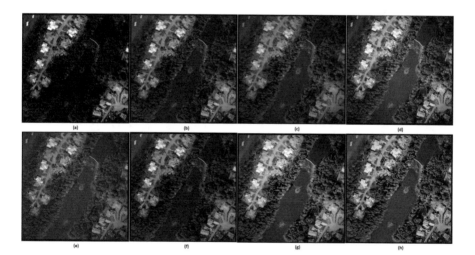

Fig. 8 Simulation results of the source PAN QuickBird image (QuickBird, 0.7-m resolution, 512×512 pixels after resizing) and enhanced results of different methods. (**a**) Original "PAN image", (**b**) GA-based technique, (**c**) DE-based technique, (**d**) ABC technique, (**e**) PSO-based technique, (**f**) ACO-based technique, (**g**) CSO-based technique, and (**h**) the proposed technique (TCO)

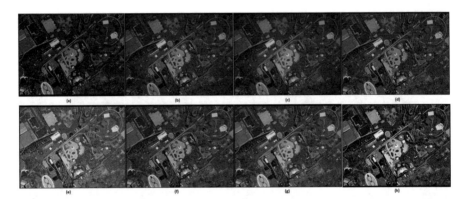

Fig. 9 Simulation results of the source PAN QuickBird image (QuickBird, 0.7-m resolution, 512× 512 pixels after resizing) and enhanced results of different methods. (**a**) Original "PAN image", (**b**) GA-based technique, (**c**) DE-based technique, (**d**) ABC technique, (**e**) PSO-based technique, (**f**) ACO-based technique, (**g**) CSO-based technique, and (**h**) the proposed technique (TCO)

9(a), we have shown different low-contrast ground truth PAN QuickBird satellite images which are taken from http://earthobservatory.nasa.gov/. These images have been enhanced using GA, DE, ABC, PSO, ACO, and CSO techniques, and the proposed TCO method. In experiments, the performance of these techniques are assessed by considering the consistency of the reconstructed enhanced image from the original image. For subjective evaluation, mean square error (MSE) [15, 19, 36], peak signal to noise ratio (PSNR) [15, 19, 36], picture quality scale (PQS) [17, 22, 28], the universal quality index (UQI) [17, 22, 28, 36], the linear index of fuzziness (LIF) [17, 28, 36], structural similarity metrics (SSIM) [22, 25, 28, 36], and assessment metrics are taken into account. In the testing stage, several satellite images are included to validate the usefulness of our algorithm. The performance of this technique is assessed in terms of the following significant metrics [17, 22, 25, 28, 36]:

$$\textbf{Mean}(\mu) = \frac{1}{MN} \sum_{x=1}^{M-1} \sum_{y=1}^{N-1} \mathbf{I}(x, y) \tag{24}$$

$$\textbf{STD}(\sigma) = \sqrt{\frac{1}{MN} \sum_{x=1}^{M-1} \sum_{y=1}^{N-1} [\mathbf{I}(x, y) - \mu]^2} \tag{25}$$

Mean (μ) is the average of all intensity values. It denotes the average brightness of the image, whereas standard deviation is the deviation of intensity values about the mean. It denotes the average contrast of the image. Here $I(x, y)$ is the intensity value of the pixel (x, y), and (M, N) is the dimension of the image. At first, the performance of the proposed algorithm is applied on multi-spectral PAN satellite sample images. Thereafter, comparison of the proposed TCO technique is conducted with the most popular metaheuristic based approaches such as GA, DE, ABC, PSO, ACO, and CSO, which shows the superiority of the proposed technique in the image enhancing problem.

In order to performance analysis, the PSNR block computes the PSNR between two images in decibels. This ratio is frequently used as a superiority measurement between the original and enhanced images. A higher PSNR indicates a better quality of the enriched or reconstructed image. MSE and PSNR are the two error metrics measured against the enhanced quality of the image. MSE represents the cumulative squared error between the enhanced and original image, whereas PSNR represents a measure of the peak error. The lower value of MSE represents a lower error. In addition, other popular quality metrics (UQI, LIF, and SSIM) are used and discussed in the subsequent section in detail. In our experiment, we do not depict the mean (μ) and variance (σ) in the case of comparison. Here, we used μ, σ to define more powerful metrics such as UQI, LIF, and SSIM respectively.

Experimentally, to provide the objective performance comparison, several standard measurement metrics (image quality measurements) such as **MSE** [9], **PSNR**, **PQS**, index **UQI**, and **SSIM** are successfully utilized in detail. They are defined as follows [22, 25, 28, 36]:

1. **MSE**: Let $x_{i,j}$ and $y_{i,j}$ denote the pixels at the position (i, j) of the original image and the reconstructed image with a size of $M \times N$ each, respectively. **MSE** is calculated by the following equation:

$$\mathbf{MSE} = \frac{\sum_{i=1}^{M} \sum_{j=1}^{N} \left(x_{i,j} - y_{i,j}\right)^2}{M \times N} \qquad (26)$$

Here, **M** and **N** are the number of rows and columns in the input images, respectively. **x** is the original image and **y** is the enhanced image.

2. **PSNR**: To compute **PSNR**, the **MSE** is used, which is defined by:

$$\mathbf{PSNR} = 10 \cdot \log_{10} \frac{Q^2}{\frac{1}{M \times N} \sum_{i=1}^{M} \sum_{j=1}^{N} \left(x_{i,j} - y_{i,j}\right)^2} \qquad (27)$$

where $Q = 255$ for 8-bit 2D images. Here, M and N are the number of rows and columns in the input images, respectively. I_1 is the original image and I_2 is the enhanced image.

3. **PQS**: PQS is calculated based on the biased difference between image X and Y, where the function $f(\cdot)$ is the frequency-selective property of the human visual system. Typically, PQS is represented in the following form [25, 36]:

$$\mathbf{PQS} = 10 \cdot \log_{10} \frac{1}{\frac{1}{M \times N} \sum_{i=1}^{M} \sum_{j=1}^{N} \left(x_{i,j} - y_{i,j}\right)^2} \qquad (28)$$

where the function $f(\cdot)$ is defined the same as in [25, 36]. A larger PQS value specifies a better image quality.

4. **UQI**: Typically, **UQI** measures the amount of image distortion in terms of three parameters: loss of correction, luminance variation, and contrast distortion. Its highest value is 1, and as being close to the maximum indicates minimum distortion. **UQI** is defined as [17, 22]:

$$\mathbf{UQI} = \frac{4\sigma_{xy}\mu_x\mu_y}{(\sigma_x^2 + \sigma_y^2)(\mu_x^2 + \mu_y^2)} \tag{29}$$

$$\sigma_{xy} = \sum_{i=1}^{M}\sum_{j=1}^{N}(x_{ij} - \mu_x)(y_{ij} - \mu_x) \tag{30}$$

where x and y are the source and final images, with its mean as μ_x, μ_y, and its variance as σ_x σ_y, respectively.

5. **LIF** [22, 25, 28, 36]: LIF is a spatial information based performance evaluation metric that is widely used to measure the quality of a processed image (especially image enhancement). For an image **I** with size $M \times N$, it is defined and written as:

$$\mathbf{LIF} = \frac{2}{M \times N}\sum_{i=1}^{M}\sum_{j=1}^{N}\min\{\mu_{ij}, (1 - \mu_{ij})\} \tag{31}$$

$$\mu_{ij} = \sin\left[\frac{\pi}{2}\left(1 - \frac{I_{i,j}}{I_{\max}}\right)\right] \tag{32}$$

where $\mathbf{I}_{ij}$ denotes the pixel intensity at spatial location (i, j), and $\mathbf{I}_{\max}$ is the maximum intensity value of the given image. A small value of LIF indicates better enhancing performance of the tested method [25, 28].

6. **SSIM** [22, 25, 28, 36]: Let, x and y be the source and final images with its mean as μ_x, μ_y, and variance as σ_x σ_y, respectively. The structural similarity index (SSIM) is defined by intern = ms of several values such as $p_1 = 0.01$, $p_2 = 0.03$, and **L** as the dynamic range of pixel intensity (for an eight-bit image $L \in [0, 255]$) as follows:

$$\mathbf{SSIM} = \frac{(2\sigma_{xy} + d_2)(2\mu_x\mu_y + d_1)}{(\sigma_x^2 + \sigma_y^2 + d_2)(\mu_x^2 + \mu_y^2 + d_1)} \tag{33}$$

where $d_1 = (p_1 \times L)^2$, $d_2 = (p_2 \times L)^2$, and σ_{xy} are defined in **UQI**.

Tables 1, 2, 3, 4, 5, and 6 show the various results of the quantitative performances of all enhanced methods for the selected images which are shown in Figs. 4, 5, 6, 7, 8, and 9, respectively. From these tables and figures, it can be seen that the proposed algorithm always achieves the highest values of all performance metrics (MSE, PSNR, PQS, UQI, LIF, and SSIM) which are listed as bold than from the aforementioned algorithms.

Table 1 Performance comparison of different methods with several metrics on the source image 4

Methods	MSE	PSNR (dB)	PQS (dB)	UQI	LIF	SSIM
GA	5.0573	42.8571	26.0570	0.6531	0.4228	0.8842
DE	4.7501	42.9715	26.1309	0.6618	0.3904	0.8931
ABC	4.4120	43.2916	26.1673	0.6897	0.3751	0.8978
PSO	4.3431	43.9881	26.2202	0.7197	0.3541	0.9904
ACO	4.2523	44.1223	26.3519	0.7262	0.3136	0.9917
CSO	4.1519	44.3739	26.5389	0.7368	0.3107	0.9922
TCO	**4.0361**	**44.9361**	**26.5961**	**0.7401**	**0.3094**	**0.9931**

Table 2 Performance comparison of different methods with several metrics on the source image 5

Methods	MSE	PSNR (dB)	PQS (dB)	UQI	LIF	SSIM
GA	4.6157	42.3185	25.1575	0.6507	0.3639	0.8023
DE	4.4167	42.3620	25.1630	0.6564	0.3463	0.8246
ABC	4.3431	43.3722	25.2206	0.7218	0.3056	0.8509
PSO	4.2122	43.4876	25.2418	0.7871	0.2840	0.8607
ACO	4.2523	44.5127	25.3059	0.7962	0.2602	0.8878
CSO	4.1416	44.7362	25.3379	0.8302	0.2505	0.8937
TCO	**4.0907**	**44.9530**	**25.3728**	**0.8540**	**0.2133**	**0.9012**

Table 3 Performance comparison of different methods with several metrics on the source image 6

Methods	MSE	PSNR (dB)	PQS (dB)	UQI	LIF	SSIM
GA	4.2772	42.9480	26.1012	0.5443	0.4409	0.8254
DE	4.2314	42.7296	26.1773	0.5820	0.4177	0.8473
ABC	4.1861	43.0970	26.2257	0.6169	0.4066	0.8566
PSO	4.1740	43.1419	26.2808	0.6305	0.3933	0.8793
ACO	4.1116	44.2711	26.3004	0.6757	0.3834	0.8935
CSO	4.0910	44.3611	26.3259	0.7099	0.3603	0.9077
TCO	**4.0082**	**44.4068**	**26.3469**	**0.7514**	**0.3263**	**0.9105**

Table 4 Performance comparison of different methods with several metrics on the source image 7

Methods	MSE	PSNR (dB)	PQS (dB)	UQI	LIF	SSIM
GA	5.6679	43.1773	27.0627	0.7531	0.5228	0.7842
DE	5.6531	43.1871	27.1080	0.7618	0.5904	0.7931
ABC	5.6268	43.1797	27.1673	0.7897	0.5751	0.7978
PSO	5.6093	43.1801	27.2202	0.7197	0.4541	0.8904
ACO	5.5855	44.0923	26.3519	0.8262	0.4136	0.8917
CSO	5.5749	44.1095	26.5389	0.8368	0.4107	0.8922
TCO	**5.5528**	**44.1133**	**26.5961**	**0.8401**	**0.4094**	**0.9931**

Table 5 Performance comparison of different methods with several metrics on the source image 8

Methods	MSE	PSNR (dB)	PQS (dB)	UQI	LIF	SSIM
GA	5.1183	43.1829	27.3945	0.8156	0.5991	0.8151
DE	5.0837	43.1972	27.12873	0.8288	0.5751	0.8404
ABC	5.0675	43.2159	27.1759	0.8361	0.5517	0.8645
PSO	4.9168	43.2584	27.0266	0.8599	0.5358	0.8941
ACO	4.6156	44.1579	26.8618	0.8769	0.5358	0.9276
CSO	4.3956	44.2832	26.3938	0.8991	0.5160	0.9431
TCO	**4.3754**	**44.3014**	**26.1753**	**0.9068**	**0.5082**	**0.9671**

Table 6 Performance comparison of different methods with several metrics on the source image 9

Methods	MSE	PSNR (dB)	PQS (dB)	UQI	LIF	SSIM
GA	4.4214	42.1649	26.3831	0.7749	0.4322	0.8655
DE	4.4185	42.2522	26.5759	0.7827	0.4129	0.8728
ABC	4.4058	43.5976	26.6719	0.8025	0.4027	0.8828
PSO	4.3924	43.7468	26.8414	0.8347	0.3910	0.8929
ACO	4.3899	44.0503	26.9072	0.8691	0.3861	0.9025
CSO	4.3680	44.1616	26.9287	0.8875	0.3773	0.9196
TCO	**4.3467**	**44.3130**	**26.9871**	**0.8964**	**0.3688**	**0.9366**

Because of the complexity problem (resolution variation, spectral preservation) in satellite image processing, most of the existing enhancement techniques have been tested on an average remote sensing dataset, or otherwise simply on gray tone images. In the case of satellite images, the rate of variation of the resolution and image detailed are certainly high because most satellite images are typically very dense. By reason of the highly compact spectral bands, the rate of variation from one image region to another region of the satellite image is really fast. Thus, for dim satellite image enhancement, exact enhancement seems to be challenging. There are several approaches, which have been developed and successfully applied to resolve enhancement problems with satellite images.

In this work, six metaheuristic approaches—GA, DE, ABC, PSO, ACO, and CSO techniques—were applied for assessment and investigation. Figs. 4, 5, 6, 7, 8, and 9 are shown the low contrast original PAN satellite images, respectively. These images were enhanced using GA, DE, ABC, PSO, ACO, CSO, and the proposed algorithm, which is shown in Figs. 4, 5, 6, 7, 8, and 9. The TCO technique is included in Figs. 4, 5, 6, 7, 8, and 9h, which provides better enhancement of the image than GA, DE, ABC, PSO, ACO and, CSO, which are shown in Figs. 4, 5, 6, 7, 8, and 9f. All the results shown in Figs. 4, 5, 6, 7, 8, and 9h represent the improved images (contrast enhanced) by the proposed TCO approach. Images obtained by remaining methods are shown in Figs. 4, 5, 6, 7, 8, and 9g. The quality of the enhanced image and improved results indicate that the proposed enhancement technique TCO can produce a sharp and brighter image than the conventional GA,

DE, ABC, PSO, ACO, and CSO approaches. In Figs. 4, 5, 6, 7, 8, and 9h, we notice that all the resulting images made by GA, DE, ABC, PSO, ACO, and CSO techniques are comparably different from the image obtained by the proposed TCO method, though these six approaches are mainly considered for validation purposes and tests have been carried out on more than 100 particular PAN QuickBird satellite images. In addition, GA, DE, ABC, PSO, ACO, and CSO techniques have great ability to solve the low contrast image enhancing problem while they meet the proper criteria (such as exact parameter setting, number of iterations, selection, and choosing a suitable global function).

Figure 4a–h displays the resulting image of TCO on the remote sensing PAN source image and when compared with the conventional GA, DE, ABC, PSO, ACO, and CSO individually. Figure 4a shows the original low contrast PAN image with resolution (512×512). The GA, DE, ABC, PSO, and ACO methods do not successfully preserve the edge information, such as for the tree. Also, it can be observed in Fig. 4b–g that image objects (tree, road, and house) in the background of the image appear slightly hazy and dim. The image quality is significantly better in Fig. 4h according to the TCO approach. Moreover, we may notice that the details of the images are clearly preserved and the variations in the shade of the background are more visible, which facilitates better human perception, although CSO obtains almost the same results as TCO, which is illustrated in Fig. 4g. Observably, Fig. 4h makes better color consistency by TCO, which is better than GA, DE, ABC, PSO, ACO, and CSO in Fig. 4b–g.

In the same way, for the following test image sets, Fig. 9a–g provides the assessment results of all algorithms with the original source PAN image of size 512×512 that cover the islands on the river under an urban scene. The houses and trees on the islands are more visible in Fig. 4h, generated by TCO. The image in Fig. 4b–f loses more detail and has reduced clarity due to improper parameter assignment. In Fig. 4g, CSO demonstrates better improved enhancement results than GA, DE, ABC, PSO, and ACO, except TCO. In contrast, TCO produces highly improved contrast with certain iterations as shown in Fig. 9g. On the other hand, the following experiments were performed on a remote sensing QuickBird satellite PAN image; the original image in Fig. 8a was taken in a part of Tokyo city with dim light. Accordingly, we may notice that the city buildings and the road seem to be slightly visible after enhancement by the standard GA, DE, ABC, PSO, and ACO in Fig. 8b–f. Figure 8g, h shows CSO and TCO offer better enhanced results than GA, DE, ABC, PSO, and ACO but not TCO provides more visible appearance than CSO. Similarly, the sharpness, better detail, and stable clarity for TCO and CSO are more consistent than for GA, DE, ABC, PSO, and ACO.

In the following assessment, Fig. 7b–g provides the expected resulting images of all considered enhancement algorithms with the 512×512 resolution QuickBird satellite PAN image which shows the buildings with connected roads in the city scene. The houses and the road appear sharper and clearer in Fig. 7g achieved by our algorithm; the background appears clearer than the source image as well as the other resultant images. The resulting images in Fig. 7b–f obtained by GA, DE, ABC, PSO, and ACO lose certain image detail and a reduced contrast level, whereas CSO and

TCO provide improved image details and an enhanced contrast level in the resulting images in Fig. 7g and h, respectively. In contrast, TCO yields an optimal contrast level with better image detail which is shown in Fig. 7f. Thus, we may remark that TCO and also CSO can modify the dynamic intensity level of the source images better than the remaining methods.

In Fig. 7g, the house beside the road is more perceivable by the TCO algorithm. In the next simulation of a QuickBird PAN image with size 512×512, all the resulting images are shown in Fig. 6b–g. In Fig. 6b–f, other comparative methods are shown to be definitely losing image detail. CSO and TCO exhibit better results in Fig. 6g, h. Then again, we notice that TCO produces the finest sharpness and with great detail in Fig. 6h, more than 6g by the CSO method. It is obvious that our method is more suitable for enhancing a dim PAN satellite scene and effectively preserving image detail than is the case with most prevailing approaches. In brief, Tables 1, 2, 3, 4, and 5 all provide the evaluated results of all comparative algorithms for all fixed size (512×512) benchmark images in terms of all metrics (MSE, PSNR, PQS, UQI, LIF, and SSIM). According to the results, our proposed algorithm demonstrates better performance in the six low contrast scenes, although GA, DE, ABC, PSO, ACO, and CSO need to adjust their scale parameters and iterations with respect to the variety of environments to attain improved results. In this chapter, the proposed model TCO utilizes an entropy global optimization function to achieve the optimal outcome by adjusting the fitness values. In the case of other methods selected for comparison, we used entropy as the global function and adopted it to their corresponding parameters. In this study, we restricted ourselves to design and implementation and compared our proposed method with most conventional metaheuristic approaches, such as GA, DE, ABC, PSO, ACO, and CSO.

To address the computational efficiency of all enhanced techniques, the execution times of the standard GA, DE, ABC, PSO, ACO, CSO, and TCO are verified and shown in Table 6 with the minute as the unit, although, due to the several parameters and different iterations, GA, DE, ABC, PSO, ACO, and CSO perform with different running times. In particular, GA, DE, ABC, PSO, and ACO take several minutes to handle an image, whereas CSO and TCO require apparently less time than for the same image. TCO achieves the best indices compared to other approaches as shown in Tables 1, 2, 3, 4, 5, and 6. On the other hand, Table 7 prominently shows the execution times of GA, DE, ABC, PSO, ACO, CSO, and TCO in minutes. It

Table 7 Comparison of execution times of the different methods in minutes

Methods	Image-4	Image-5	Image-6	Image-7	Image-8	Image-9	**Average**
GA	1.2859	1.2855	1.2865	1.2854	1.2856	1.2851	1.2856
DE	1.2858	1.2859	1.2863	1.2857	1.2858	1.2853	1.2858
ABC	1.2859	1.2854	1.2864	1.2856	1.2857	1.2850	1.2856
PSO	1.2857	1.2859	1.2867	1.2853	1.2855	1.2855	1.2857
ACO	1.2855	1.2856	1.2862	1.2855	1.2859	1.2852	1.2857
CSO	1.2853	1.2858	1.2863	1.2854	1.2853	1.2854	1.2855
TCO	**1.2851**	**1.2856**	**1.2862**	**1.2853**	**1.2849**	**1.2849**	**1.2854**

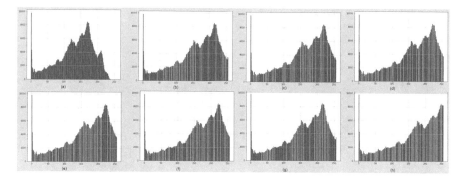

Fig. 10 Histogram comparison of different enhancement methods of the original source image 4. (**a**) Histogram of the original image, (**b**) Histogram of the GA technique, (**c**) Histogram of the DE technique, (**d**) Histogram of the ABC technique, (**e**) Histogram of the PSO technique, (**f**) Histogram of the ACO technique, (**g**) Histogram of the CSO technique, and (**h**) Histogram of the proposed technique (TCO)

demonstrates that the TCO algorithm can minimize the computational complexity within a specified iteration.

In order to validate the performance of the proposed TCO method, we performed not only a subjective evaluation but also constructed histograms for comparison. We generated the histogram of each result of the above mentioned algorithms with an image set Fig. 4. We noticed that the histogram of the original source PAN image in Fig. 10a is more narrow within the dynamic gray level range [0–256], whereas each and every histogram of the comparison methods used (i.e. GA, DE, ABC, PSO, ACO, and CSO, as shown in Fig. 10b–h), except the proposed method, modify the dynamic gray level range. However, the histogram of TCO, as shown in Fig. 10h, reveals that TCO is able to modify the dynamic gray level range more than the others, due to its better enhancing ability.

In this section, we show all resultant images generated by different methods, that is GA, DE, ABC, PSO, ACO, CSO, and TCO, and present all the above mentioned performance evaluation matrix values in Tables 1, 2, 3, 4, 5, and 6 with selected sets of benchmark images. However, due to page constraints, we only present the graphical plots for the image sets Fig. 4. Here, we show all performance comparisons in terms of **bar chart**, **pie chart**, and **line plots**. For example, Fig. 11a–d illustrate all the matrix values provided by different methods used on image sets, as shown in Fig. 4 with **bar chart**, **pie chart**, and **line plots**. On the other hand, the run time for individual methods for all image sets is presented in Table 7 and, for better understanding, we represent the obtained result of Table 7 by **bar chart**, **pie chart**, and **group plot**, separately. Average run times of all the above mentioned methods for all image sets are measured and shown in Table 7 with bold notation and by Figs. 12 (a), 12 (b) and 12 (c) are bar plot (average runtime), pie plot, and group bar plot (different runtime) which are illustrated the runtimes of all methods, respectively. On the basis of all these methods and benchmark images, we may

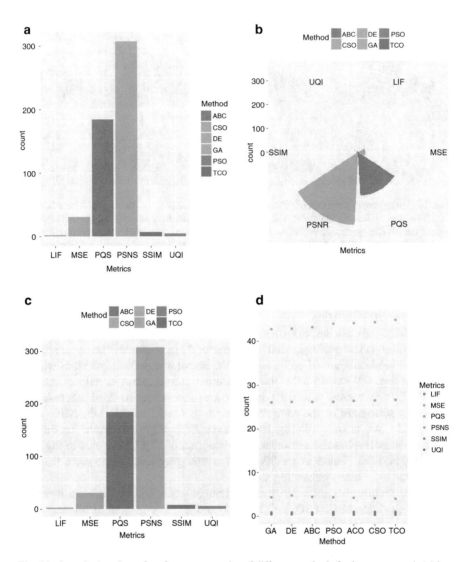

Fig. 11 Quantitative plots of performance metrics of different methods for image group 4: (**a**) bar plot, (**b**) point line plot, (**c**) pie plot, (**d**) line plot

greatly expect that the proposed method would be an alternative image enhancing method for low contrast images in the domain of remote sensing.

In order to establish the effectiveness of TCO, estimated data in terms of several matrices of MSE, PSNR, PQS, UQI, LIF, and SSIM are presented in Tables 1, 2, 3, 4, and 5 and are truly compared with the performances for GA, DE, ABC, PSO, ACO, CSO, and TCO to specify the validation of our experiments. Furthermore, in order to make a complete evaluation, all methods are executed with

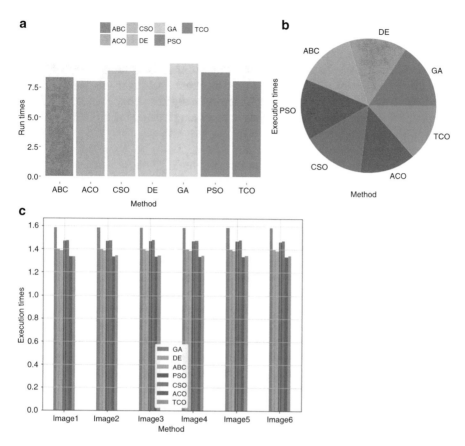

Fig. 12 Quantitative plots of overall run time of different methods: (**a**) bar plot(average), (**b**) pie plot, (**c**) group bar plot

all the used benchmark PAN satellite images and their processed times are recorded to understand their performance. The results in Tables 1, 2, 3, 4, and 5 indicate that the proposed technique (TCO) over-performs the above techniques in enhancing image contrast.

The primary advantage of the proposed TCO technique is that it is adept at enhancing the poor and low-contrast satellite PAN images when devised for very low illumination. The metaheuristic optimization enhanced techniques can perform better if their corresponding parameters are optimally set. However, if the input PAN image contains very dim images and the variation of the pixel intensity in the image is fixed then these techniques are not able to perform better enhancement. As shown in Table 7, the proposed technique has less execution time than the others where the input satellite images have a fixed resolution (512 × 512). Therefore, Tables 1, 2, 3, 4, 5, and 6 demonstrate the superiority of the proposed algorithm especially for very low and poor-contrast PAN satellite images.

In this work, only a large collection of PAN QuickBird satellite images are considered and applied to test performance validation of our proposed enhancement algorithm. We have compared our approach with the existing most popular meta-heuristic optimization based enhancement approaches for comparative (subjective and objective assessment) studies. From all these facts, we see that the TCO approach not only enhances the input images visually but also produces sharper and brighter enhanced images which assists in producing a better interpretation than the state-of-the-art satellite image enhancement approach.

From these plots of Figs. 11a–d and 12a–c, it can be observed that for all benchmark low-contrast PAN satellite images, the TCO performed a considerable improvement in contrast enhancement. From this comparative plot of MSE, PSNR, PQS, UQI, LIF, and SSIM matrices, it is clear that the proposed method provides better enhancing performance in low contrast PAN images. Similarly, statistically, the performance analysis in terms of PQS, UQI, LIF, and SSIM are plotted in Fig. 11 for different images with different approaches. From the plots in Fig. 11a–d based on metric records, it is clear that there is an improvement found in edge sharpness and clarity by the proposed TCO approach compared to the other approaches. According to the all data plots of the aforesaide performance matrices, it can be observed that the assessment values of PQS, UQI, LIF, and SSIM are highest and it is achieved by the proposed approach TCO which are listed in bold in the all tables. Therefore, it is clear that the TCO can make a significant improvement in image enhancing task and it can be more suitable for real-time remote sensing applications, specifically low light contrast improvement. Finally, the plots shown in Fig. 12 depict a considerable improvement in the execution time by the proposed method.

From all points of view, to establish a true validation of TCO for enhancing the PAN satellite image, in terms of several performance measures, metrics like MSE, PSNR, PQS, UQI, LIF, and SSIM are carefully considered and reported in Tables 1, 2, 3, 4, 5, and 6 individually. Based on the values of these indices, it is validated that the proposed TCO approach not only outperformed other approaches but also successfully enhanced the low contrast PAN image in a reasonable run time. In the plots of Figs. 11a–d and 12a–c, it has been established that there is not only an improvement in contrast but also an improvement in image detail by the proposed approach with an effective cost of execution.

5 Conclusion

We have proposed a TCO algorithm in this work and successfully utilized it for contrast enhancement in low light or unevenly illuminated images. This TCO algorithm along with fuzzy type-2 methodology has been applied to the image enhancement technique on low contrast color images. To optimize the informational quantity in an enhanced image, we operated a fitness function such as fuzziness in the fuzzy domain. The fuzzifier and intensification parameters were evaluated automatically for the input color image of the TCO system in the fuzzy domain.

This technique has been successfully tested on various low contrast color images. Experimental results show the superiority of our proposed technique over prevailing conventional methodologies.

References

1. S.S. Agaian, B. Silver, K.A. Panetta, Transform coefficient histogram-based image enhancement algorithms using contrast entropy. IEEE Trans. Image Process. **16**(3), 741–758 (2007)
2. B. Akay, A study on particle swarm optimization and artificial bee colony algorithms for multilevel thresholding. Appl. Soft Comput. **13**(6), 3066–3091 (2013)
3. T. Arici, S. Dikbas, Y. Altunbasak, A histogram modification framework and its application for image contrast enhancement. IEEE Trans. Image Process. **18**(9), 1921–1935 (2009)
4. S. Bakhtiari, S. Agaian, M. Jamshidi, A color image enhancement method based on Ensemble Empirical Mode Decomposition and Genetic Algorithm, in *World Automation Congress (WAC)* (IEEE, Piscataway, 2012), pp. 1–6
5. A.K. Bhandari, V. Soni, A. Kumar, G.K. Singh, Artificial Bee Colony-based satellite image contrast and brightness enhancement technique using DWT-SVD. Int. J. Remote Sens. **35**(5), 1601–1624 (2014)
6. A.K. Bhandari, V. Soni, A. Kumar, G.K. Singh, Cuckoo search algorithm based satellite image contrast and brightness enhancement using DWT-SVD. ISA Trans. **53**(4),1286–1296 (2014)
7. G.G. Bhutada, R.S. Anand, S.C. Saxena, Edge preserved image enhancement using adaptive fusion of images denoised by wavelet and curvelet transform. Digital Signal Process. **21**, 118–130 (2011)
8. J. Chen, W. Yu, J. Ti, Image contrast enhancement using an artificial bee colony algorithm. Swarm Evol. Comput. **38**, 287–294 (2017)
9. H.D. Cheng, H. Xu, A novel fuzzy logic approach to contrast enhancement. Pattern Recogn. **33**(5), 809–819 (2000)
10. E. Daniel, J. Anitha, Optimum wavelet based masking for the contrast enhancement of medical images using enhanced cuckoo search algorithm. Comput. Biol. Med. **71**, 149–155 (2016)
11. H. Demirel, G. Anbarjafari, Discrete wavelet transform-based satellite image resolution enhancement. IEEE Trans. Geosci. Remote Sens. **49**(6), 1997–2004 (2011)
12. H. Demirel, C. Ozcinar, G. Anbarjafari, Satellite image contrast enhancement using discrete wavelet transform and singular value decomposition. IEEE Geosci. Remote Sens. Lett. **7**(2), 333–337 (2010)
13. C. Dong, C.C. Loy, K. He, X. Tang, Learning a deep convolutional network for image super-resolution, in *ECCV* (2014)
14. A. Draa, A. Bouaziz, An artificial bee colony algorithm for image contrast enhancement. Swarm Evol. Comput. **16**, 69–84 (2014)
15. Z. Fan, D. Bi, W. Ding, Infrared image enhancement with learned features. Infrared Phys. Technol. **86**, 44–51 (2017)
16. R.C. Gonzalez, R.E. Woods, S.L. Eddins, *Digital Image Processing* (Addison-Wesley, Boston, 1992), pp. 127–211
17. M. Hanmandlu, O.P. Verma, N.K. Kumar, M. Kulkarni, A novel optimal fuzzy system for color image enhancement using bacterial foraging. IEEE Trans. Instrum. Meas. **58**(8), 2867–2879 (2009)
18. S. Ioffe, C. Szegedy, Batch normalization: accelerating deep network training by reducing internal covariate shift, in *ICML* (2015)
19. N.H. Kapla, Remote sensing image enhancement using hazy image model. Optik-Int. J. Light Electron Optics **155**, 139–148 (2018)

20. S. Lee, H. Kwon, H. Han, G. Lee, B. Kang, A space-variant luminance map based color image enhancement. IEEE Trans. Consum. Electron. **56**(4), 332–338 (2010)
21. C. Li, Y. Yang, L. Xiao, A novel image enhancement method using fuzzy Sure entropy. Neurocomputing **215**, 196–211 (2016)
22. S.H. Malik, T.A. Lone, Comparative study of digital image enhancement approaches, in *Computer Communication and Informatics (ICCCI)* (IEEE, Piscataway, 2014), pp. 1–5
23. L. Maurya, P.K. Mahapatra, A. Kumar, A social spider optimized image fusion approach for contrast enhancement and brightness preservation. Appl. Soft Comput. **52**, 575–592 (2017)
24. G. Michal, C. Gaurav, P. Sylvain, D. Frdo, Deep joint demosaicking and denoising. ACM Trans. Graph. **35**, 1–12 (2016)
25. K. Narasimhan, V. Elamaran, S. Kumar, K. Sharma, P.R. Abhishek, Comparison of satellite image enhancement techniques in wavelet domain. Res. J. Appl. Sci. Eng. Technol. **4**(24), 5492–5496 (2012)
26. A. Polesel, G. Ramponi, V.J. Mathews, Image enhancement via adaptive unsharp masking. IEEE Trans. Image Process. **9**(3), 505–510 (2000)
27. E.H. Ruspini, Numerical methods for fuzzy clustering. Inf. Sci. **2**, 319–350 (1970)
28. H.K. Sawant, M. Deore, A comprehensive review of image enhancement techniques. Int. J. Comput. Technol. Electron. Eng. **1**(2), 39–44 (2010)
29. P. Shanmugavadivu, K. Balasubramanian, Particle swarm optimized multi-objective histogram equalization for image enhancement. Opt. Laser Technol. **57**, 243–251 (2014)
30. V. Soni, A.K. Bhandari, A. Kumar, G.K. Singh, Improved sub-band adaptive thresholding function for denoising of satellite image based on evolutionary algorithms. IET Signal Process. **7**(8), 720–730 (2013)
31. C.C. Sun, S.J. Ruan, M.C. Shie, T.W. Pai, Dynamic contrast enhancement based on histogram specification. IEEE Trans. Consum. Electron. **51**(4), 1300–1305 (2005)
32. H.H. Tsai, Y.J. Jhuang, Y.S. Lai, An SVD based image watermarking in wavelet domain using SVR and TCO. Appl. Soft Comput. **12**(8), 2442–2453 (2012)
33. O.P. Verma, P. Kumar, M. Hanmandlu, S. Chhabra, High dynamic range optimal fuzzy color image enhancement using artificial ant colony system. Appl. Soft Comput. **12**(1), 394–404 (2012)
34. X. Wang, L. Chen, An effective histogram modification scheme for image contrast enhancement. Signal Process. Image Commun. **58**, 187–198 (2017)
35. H.T. Wu, J.W. Huang, Y.Q. Shi, A reversible data hiding method with contrast enhancement for medical images. J. Vis. Commun. Image Represent. **31**, 146–153 (2015)
36. D. Yugandhar, S. Nayak, A comparative study of evolutionary based optimization algorithms on image quality enhancement. Int. J. Appl. Eng. Res. **10**(15), 35247–35252 (2015)
37. C. Zhou, H.B. Gao, L. Gao, W.G. Zhang, Particle swarm optimization (PSO) algorithm. Appl. Res. Comput. **12**, 7–11 (2003)

Image Segmentation Using Metaheuristic-Based Deformable Models

B. K. Tripathy, T. R. Sooraj, and R. K. Mohanty

Abstract The goal of the segmentation techniques called deformable models is to adapt a curve in order to optimize the overlapping with another image of interest with the actual contour. Some of the problems existing in optimization involve choosing an optimization method, selecting parameters, and initializing the curve. All these problems will be discussed within this chapter, with reference to metaheuristics, and are designed to solve complex optimization and machine learning problems. We discuss image segmentation techniques which depend on active contour models using metaheuristics. Similarly, histological image segmentation techniques are elaborated using a level set approach based upon metaheuristics.

Keywords Metaheuristics · Active contour models · Histological images · Level set

1 Introduction

In computer science and electronics, image segmentation is one of the growing fields. It has wide applicability in the medical field, satellite image processing, and so on. The main problem that we face is the presence of noise in the images. The quality of the existing image segmentation techniques like edge detection and thresholding becomes reduced in the presence of noise. Curves and surfaces come under deformable models (DMs) when these are defined within a flexible image domain, where the flexibility is provided by different forces acting externally and internally. The design of internal forces is made in such a way that the deformation process applied to curves or surfaces are achieved smoothly. On the other hand the

B. K. Tripathy (✉) · R. K. Mohanty
VIT University, Vellore, Tamil Nadu, India
e-mail: tripathybk@vit.ac.in

T. R. Sooraj
Department of Computer Science, Providence College of Engineering, Chengannur, Kerala, India

© Springer International Publishing AG, part of Springer Nature 2018
S. Bhattacharyya (ed.), *Hybrid Metaheuristics for Image Analysis*,
https://doi.org/10.1007/978-3-319-77625-5_6

external forces move the model towards the object boundary and are computed from the image data. The DMs are divided into two basic categories:

1. Parametric deformable models (PDMs)
2. Geometric deformable models (GDMs)

Under Category 1, as the name suggests, parametric forms of the curves and surfaces are used during the process of deformation. The advantage of parametric representation is that it is compact and the real time application thereafter can be fast. In the case of PDMs, aggregating and separating different parts is difficult. For GDMs, the handling of changes in structure is very natural. There exist different models which are based on curve evolution [6], the level set method, active contour methods, and so on. An implicit representation technique is followed for presenting curve evolution and a level set of functions of higher dimension non-vector functions. First a complete deformation is carried out and then an evaluation of the parametrization, as in this process accommodation of topological adaptivity is done easily. Identical principles are followed for the two methods.

Optimization algorithms are classified into two categories

1. Exact algorithms: These are the ones in which optimal solutions are obtained in a finite amount of time.
2. Heuristic algorithms: These are the one which can find better solutions, though an optimal solution is not guaranteed.

So, depending upon the nature of the output we need, we can select either exact or heuristic algorithms. Heuristics cannot be used to find the exact solution within the stipulated time, so in those cases where exact solutions are required, then exact algorithms are the best ones to use. There is not a way to measure practically the closeness of the solution obtained through heuristics and the global optimum.

Most of the heuristic algorithms are specific and problem dependent, though on the other hand metaheuristics are problem independent. Metaheuristics have their own guidelines and strategies for developing heuristic optimization algorithms. The main difference between heuristics and metaheuristics is that the latter repeatedly try to optimize the problems through computational methods. This enables us to search in large spaces of candidate solutions.

The categorization of metaheuristics is as follows:

1. Trajectory methods (TMs): In this method the search process forms a trajectory in the search space which seems to be the generation of a discrete dynamical system in non-continuous time. In tabu search, simulated annealing, iterative local search, and variable neighborhood search, TM is used.
2. Population based methods: A population of solutions is searched during each individual iteration step of these methods. It is seen that a member of the solution space is generated in discrete time. Evolutionary algorithms (EAs) such as genetic algorithms [16, 19], evolution strategies [2], and evolutionary programming [13] come under this paradigm. It also includes particle swarm optimization (PSO) [22]. Biological evolutionary terms such as reproduction,

mutation, recombination, and selection form the basic notions of evolutionary algorithms, which are basically used to solve optimization problems. The collective behavior of decentralized, self-organized artificial systems is used in swarm intelligence methods. A more recent and popular algorithm for non-discrete optimization is differential evolution [37]. This method borrows features from EAs as well as swarm intelligence methods.

3. Memetic algorithms: These are generalizations of Genetic Algorithms (GAs). The chance of untimely convergence is controlled through local search techniques. A local enhancement method is embedded into an algorithm based upon population and is a hybridization of different search methods which may be either local or global [31]. Overall its effect can be described as the reproduction of learning and societal interaction in a life time of people getting global search solutions being effected by the optimization of their local counterparts. As an example, we can take the case of scatter search [15].

2 Image Segmentation (IS) Using Metaheuristic Based DMs

IS can be defined as the dividing of images into components such that the intersection of these components must be empty. To segment an image, one has to consider the attributes which are related to the image. Some of those attributes are image type, others are image intensity. Some of the areas relating to IS are medical images, object detection, and video surveillance. IS can be classified as thresholding techniques, edge based methods, region based approaches, and DMs. In thresholding techniques, the segmentation is based on the intensity of the pixel. In the case of edge based methods, it is based on boundary localization. If the segmentation is based on region detection, then we can call it a region based approach; and if it is based on shape, then we can call it a DM.

In this chapter, we are focused on DMs and metaheuristics and how they are related to IS. As mentioned earlier, DMs start from curves and are modified due to internal and external forces (shrinking or expansion operations). The aim of these operations is minimizing the energy function. A consequence of this is that the boundary of the object to be segmented must fit to the curve. Thus it boils down to the fact that IS can be viewed as optimizing a function which is multimodal.

The significant properties of metaheuristics have helped researchers to apply it with IS using DMs. Mainly IS techniques use metaheuristics and are applied in medical images [27, 45]. Zografos [45] used concepts in genetic algorithms which helped to improve the convergence speed. Maulik [27] discussed the suitability of various optimization techniques by conducting tests over a series of two-dimensional analytical functions. The tests conducted by them highlighted the properties of the various optimization techniques. In the next section, we discuss the metaheuristic approaches using an active contour model.

2.1 Active Contour Models (ACMs)

Kass et al. introduced a concept called snakes, which is a model for detecting the boundaries of objects coming under the general concept of ACMs. In order to minimize the energy function so that it coincides with the boundary of the object a deformation procedure is used. A latest method to achieve contour based segmentation is genetic algorithm (GA) snakes. The main features of this method are related to complexity, storage, and storage space [26].

Active contour modeling can be efficiently used in object boundary extraction. In [26] the vector bundle (VB) constraint method, based on ACMs for PSO, is discussed. This method can be differentiated in the sense that VB is predefined whereas the restriction in velocity update is its bundle. The particles in PSO are represented by the control points on the contour in order to apply the idea of ACMs; the particles drive the evolution procedure. Many research publications have come out of the interest of researchers in classic contour models in swarm intelligence [12].

Simulated annealing (SA) is a technique to approximate global optimization in a large search space. Some of the techniques using SA and basic memetic approaches are discussed in [39]. Tang et al. discussed an improved snake model using SA. This model differs from the existing models in the use of center energy in the traditional snake model; curved energy can be tuned according to the pot. This will help the initialized curve to store both topological properties and also to fit the concave of the object.

Nowadays metaheuristics are applied in IS, and one of the most useful models in this direction is (ACMs). For most of the methods, metaheuristics is used for evolving control points of the model such that in the search space best locations can be searched. Several applications of IS have evolved from the idea of geometric snakes introduced in [3]. Many different strategies which use a coarse-to-fine approach in the form of multi-scale segmentation, multi-stage evolution, or the combination of different optimization methods at different stages have been proposed.

To search for the best possible set of parametric values in generating ACMs several methods have been proposed. These parameters are basically the terms involved in defining an energy function. This is a supervised learning process where the training methods are used for finding optimal values of the parameters and where the training data are obtained from manual segmentation. GAs are used by Rousselle et al. [36] to propose a mechanism for parameter adjustment. The two approaches introduced in [36] are categorized as: a global set of parameters generated through a supervised approach, where the evaluation functions of the GA are done by using Greedy algorithms; an unsupervised approach through which local sets of parameters are generated. The GA is used to find the parameter set minimizing the energy at each point in the neighborhood of the current point in the Greedy algorithm. The noise in medical images is removed and its low quality is improved in the approach of Talebi et al. [38], where GAs are combined with ACMs. This algorithm is advantageous as the learning and optimization capabilities of GA are used such that the active contour model is evolved jointly with the weights of the energy terms. Metaheuristics have been used for setting the initial location of the snake and to calculate how many snakes are required and the appropriate number of control points. SA is used to find approximate global optimization through a metaheuristic approach in a large search space. In [35], optimization of energy function in SA is used to determine the state of neurons in a Hopfield network.

2.1.1 Encoding

There exist different methods in encoding and these approaches can be grouped on various factors, that is they may be based on the way they evolve control points or on the basis they learn energy term weights or on the way they initialize the model. But the encoding schemes used in the approaches are not the same. Normally the position of the snake is considered for encoding the images in polar coordinates and the chromosomes contain the count of the snake control. The encoding scheme is used by the authors [3–5]. Also, there are variants of this scheme for Cartesian coordinates. The other approaches also use Gray-coding or real-number encoding [42]. Some other proposals in similar circumstances follow no encoding of the control point coordinates.

MacEachern et al. [26] discussed a method for active contour optimization, where we have S_i, C_i and V_{ij}. Here C_i is the contour of the bit coded contour whose state is represented by S_i after the ith state transition, following the cardinality of the parameter set V_{ij}. The vector contour state is obtained by adding the displacement vectors to V. The detection of brain tumors and deformity boundaries for medical images is obtained by using the wavelet based pre-processing method [32]. In this method for each control point we have to encode the distance as well as the angle from the center to the control point. Obtaining the contour of an object accurately was the objective while applying GAs. The components of the internal energy are continuity energy, curvature energy, and image energy. The energy function depends upon the internal energy. In the earlier stages an active contour model was used in optimization problems. But, later many researchers applied it to derive the solutions of partial differential equations. The pros and cons of active contour based models are speed and instability respectively. So, to improve the outcome of active contour based models Fan et al. introduced a parallel GA based active contour model and used it in segmenting the lateral ventricles from magnetic resonance imaging (MRI) images. Here, a chromosome consists of three numbers of 2D arrays with the Fourier descriptors as the entries [11], and in [33] a location dependent determination of control points and their indices are used to fix a chromosome in the form of an integer array. In some other cases an identical approach is used to encode data using swarm intelligence. The search space is developed by using control points in [11], where as in [12], a swarm of particles around the control points in the snake are made to be related when searching in a neighborhood around them.

In some cases, metaheuristics are used to learn the energy term weights. Here one method differs from another due to the encoding scheme used even if they use a GA. These algorithms have the capability to learn from examples. A procedure for the contour based segmentation of 3D physical information is carried out by Cagnoni et al. [8] for evolving adaptive procedures in it. The edge detector parameter and the DM parameters are extracted from the examples. At first, a user has to determine multiple 2D contours of relevant parallel planes from the anatomical structure through division of datasets using arbitrary boundaries. The evolution of contour detectors is generated by taking examples for training in connection with GAs.

A complete segmentation of a structure of an image is obtained by using a detector on the ordered occurrence of images. Other ordering of images can be exposed to the same detector. The control of segmentation is carried out through a contour-tracking strategy which is dependent upon an elastic-contour model where GA is used to optimize the parameters. The two parameters, edge detectors, and DM are used in [42], which are characteristically very different from each other. Here, the parameters involved, including the real valued ones, are encoded by using six-bit binary strings, which transform into a genome sequence of 96 bits. Encoding of parameters for Boolean models is carried out in [7]. This process includes (1) the mapping of image values to image potentials, (2) the strength of the deformation force, (3) the appearance and intensity of gradients, (4) the minimal and maximal edge length, (5) the scale of gradients, and (6) the sign of the pressure. Individuals in this universe are parameters $Q_i = q_{i,j}^{-\!\!\!>}$, comprising the values of parameters and their variability, which is responsible for the variation of values of the population that becomes evolved.

Automatic segmentation of cardiac MRI is dealt with in [40]. This technique is mainly confined to GA to achieve optimization of the parameters used in the configuration. In the process of encoding, the three parameters of elasticity, rigidity, and viscosity are considered as components, and some additional parameters like the largest amplitude associated with the image gradient and the balloon forces are weighed to determine the strengths. Binary encoding is used to represent all these parameters. The crucial point is in the understanding of the energy terms associated with the active contour model which are used to study the corresponding nature of the objective functions for an image. An auto-learning architecture basing upon the snake model is presented in [10], which comprises the two sections associated with learning and detection. In this, the required object contour is determined by the user and weights are determined through a training method. The Taguchi approach was used by the authors for the determination of the weight ratios among energy terms. The difference between the above two methods lies in the determination of control points on the basis of local weights or global ones. The control points have weighted features. The supervised and unsupervised approaches used in [36] use a Greedy algorithm in implementing a framework. The difference between these two approaches is in the generation of parameters. The first one generates a global set of parameters and that of the second one is a local set. The GA computes a set of parameters which minimize the energy at each point in the neighborhood of the current point in the Greedy algorithm.

A two-phased GA is used in [36], comprising the first phase where the GA learns the global snake parameters and in the next phase where encoding of the generated parameters is done based upon the snake points. Continuity, curvature, gradient, intensity, and balloon force are the parameters used in [36]. Encoding by using a set of binary genes is done in [38] for continuity, curvature, image, and pressure, considered as the control coefficients of a balloon model.

There are two approaches which execute the initialization mechanism through the active control model and metaheuristics. Encoding into a binary chromosome by a Canny edge detector is performed through the threshold and sigma parameters in [38]. In the second one, initialization of the snake is carried out through swarm intelligence. Three variables, namely position, velocity, and energy, are used to define the agent's behavior.

2.1.2   Operators

Normally binary or real-coded encoding schemes are used in the implementation of operators in evolutionary algorithms. There are a lot of constraints like disallowing the crossing of links in snake models. Ad hoc operator design helped to overcome these limitations. If the fitness function [29] is used in penalty terms then we can neglect the above mentioned restrictions. There is another standard approach which penalizes unfeasible solutions with the help of a fitness function. So, uniform mutation is used in most of the methods. It is different for bit coding where flipping of bits is used.

There are some suggestions which use a scheme instead of the mutation operator. [42] proposed a bi-phase procedure for the segmentation of the left ventricle which uses landmark detection; the outcome being evolutionary snakes. This approach is widely used in angiograms in human beings. All these approaches including those of [33] as well as [25] use non-uniform mutation. This approach was first introduced by Michalewicz [30]. However, [18] used a random control point assigning technique for members of the universe, basing it upon a temporary mutation design.

A chromosome requires a number of bits to be completely encoded; the mutation rate is related to this number. One of the most popular methods applied to the mechanism of selection is the roulette wheel method [3–5]; next comes tournament selection [36]. In [26] the implementation of a rank-based selection operator is made and differentiation of inter- and intra-population selection mechanisms is obtained in [33]. In the intra-population method out of the three chromosomes selected, the selection of two parents is done on the basis of the largest average distance between the control points [44]. On the other hand, the selection of parents is done on the basis of population interest, called the reference population. The selected chromosomes are called candidate populations. Sometimes the chromosomes are selected from the reference and candidate populations. These are called the major and minor chromosomes or parents, depending upon appropriateness. The selection of a minor parent from the candidate population for crossing with a major parent is randomly selected from a candidate population. While most of the mechanisms apply to crossover operators, elitism is rarely used in a few mechanisms [42].

As a special case, it should be noted that all the binary-coded approaches use two-point [3, 36], one-point, or uniform crossover [26]. On the other hand, no specific approach is utilized for real-coded chromosomes. However, some ready-made operators are proposed without any logical support. Some of the standards in this direction are (BLX-α), which is employed in [28] and linear crossover in [42]. As for ready-made crossover techniques, a couple of such operators are implemented in [36], the selection of one among them is by a fixed number of individuals. In the first approach only one child is created from one pair of parents, for which the method of random selection is used with the procedure being selection of parameters confined to uniform distribution over a range. In the other approach the mean of the values of the parameters is computed and the single point crossover technique is used which determines the range being defined. The intra- and inter-crossovers are employed by some other authors as in the case of a selection operator [34]. The use of crossover operators is different for the two approaches. In the first approach uniform, arithmetic, and linear arithmetic crossover operators are used and in the second a majori̇parent is split and used to determine the location of the split at random. This location helps in generating a legitimate child which is obtained by selecting a segment at a randomly chosen location in the minor parent and replacing it with one of its ends.

The selection of a population initially is a region generated between two user defined radii as in [3–5]. The differentiating factor between the approach of [21] and that of Ballerini is the selection of the initial population. Also, the initial selection is a square window in the approach of Ballerini which is substantially different from a radii being selected by Hussain [21]. The sorting of individuals on the basis of the fitness function is the distinguishing feature in the approach in the GA proposed in [27] and the selection of single parents by using the Poisson distribution such that it is most likely to select the individuals who are more fit. The parameter value selection is done by taking the initial value as one-fifth of its absolute value. In another approach in [42], the selection of the first population is carried out by arbitrary deformation of the landmarks found in a given neighborhood and, for the GA, the first universe consists of the weights obtained by Taguchi's method. There are several deviations from the ones described above in the form of evolutionary algorithms which are based on swarm intelligence.

In [11] the approach is a customized PSO algorithm where the concave boundaries and local minima are not used so that it will not be sensitive to noise. The average position of particles at a point of time t is considered as an approximation of the center of mass of particles. To reduce the time of execution of the algorithm it is stopped when the snake stagnates automatically without any external force being there. To this effect the PSO equations are modified with an external energy term being included in them. Another approach is followed in [18] where the first set of values and the further motions of the particles in the swarm of particles are computed by using the boundary as the window for searching solutions. The boundary for example can be the perpendicular bisector of the line connecting two neighboring control points. Through this approach sudden changes are controlled, which in turn a snake evolution looks more appropriate by it not crossing itself. Finally, classical SA based approaches have been proposed in [1]. Sometimes a logarithmic cooling agenda is used by considering the Boltzmann distribution [43]. This approach starts with the selection of a contour in the 2D image space of dimension two such that the number of vertices is predetermined. A contour is formed for each pixel by taking a square matrix of pixel of dimension n. After that, an energy function is used to select a new vertex as candidate. For instance, the number of iterations is used to accept a vertex; further, the same technique is followed to determine the cardinality of the set of accepted vertices and whether it leads to an incorrect solution or not. Also, another property is used for elimination or creation of new vertices. The property is derived from the elements of the contour. Before the elimination or creation of a new vertex all the vertices in the contour are tested at a particular temperature. A criterion for the elimination of a vertex is derived by fixing a value as the minimum distance and those vertices are eliminated whose distances are less than the fixed minimum distance. In parallel, a maximum distance value is fixed, and if the distance between a pair of vertices is greater than this maximum distance value, a new vertex is created.

2.1.3 Fitness Function

In the common practice of using metaheuristics, the snake's energy is required to be minimized, which is a fitness function, the aim being to evolve the control points of an active contour model. Depending upon the objects to be segmented the methods change from one to another as the active contour model terms change. In fact, the external terms in [4, 42] include the gradient of the image plus a slightly different edge functional whose minima lie on the zero-crossings of $\nabla^2 G_\sigma * I(x, y)$. Through these methods the eyes' foveal avascular zone receives an additional term called the energy term. As a result, the image's energy is modified by considering the two factors associated with gradient: magnitude and direction.

The formulation developed in [4] is improved by defining a contour as a parametric curve which is piecewise continuous and uses a cubic B-spline. Further, the Lai and Chin curvature and continuity terms are added to the cubic B-spline model. Genetic snakes introduced in [3] have been used for obtaining improved solutions in different applications; like segmenting connective tissue in meat images, exploring additional internal and external energy terms, and applying them to color images [4]. The addition of a new component in the energy equation involves the gradient of the components of RGB color model; as a result change occurs in the contour directly proportional to the sign of the constant used in the formula. Another additional term in the energy equation compels the snake to contain a predefined reference area. The harmonic potential form of this term reaches the lowest value when the area enclosed by the snake is equal to the reference area. Another change is introduced into the fitness function by including a priori knowledge, so that bones in radiographic images can be segmented [5]. An identical number (three) of snakes represent the corresponding bones and the whole combination forms a binding force. It may be noted that the position of the bones are essential but not their geometrical structure. Thirty-six points form the structure of a snake and five pairs of successive points at the junctions form the binding energy. In addition to the above form, where internal and external energy terms were proposed, two more terms are added. These are (1) a derivative energy which assumes its minimal value on the edge of the image and has brighter and darker regions positioned to the left and right of the snake and (2) a model of the anatomical relationship between adjacent bones which connects suitable points of snakes close to each other. The procedure followed in [11] is the same as in [32], but there the application was different in the sense that it was to generate videos of human body components and the persons themselves. However, a change was made by removing the derivative term and adding a fresh term whereby color images can be taken care of and which involves the derivatives of each of the three components of white light, that is red, green, and blue. This technique was also used in [5].

In [32] the requirement was to apply segmentation to the brain image including both white and gray matter and the fluid in the cerebral spinal part. But the approach uses a few technical internal terms in the form of continuity and curvature and external terms in the Kirsch spatial gradient [23]. They further extended their work by changing the internal features to be dynamic during the process of training and in the later stages through slices, and exactly the opposite is done for the external features.

Contour length and center were considered to be internal features as were the intensive associations in the direction of the gradient for different contour points and the related neighbors as external features. The fitness function took care of knowledge acquired previously in the destination object, which was the mouth. A high intensity level is used for the lips; but blurring effects are seen for the teeth which is also the case for the interior of the mouth. During the Sobel filtering, a processor is used for putting into binary form and process related to morphology are applied. The objective is to place snakes on the outer and inner lip contours.

A GA is used in [42] to find out the most suitable location, which is carried out in two steps. The steps or stages are such that the first one is for internal energy while the other one is for the generation of external energy to be added to the internal one. A parallel double stage procedure is suggested which can provide solutions to two major problems; namely the response to the sensitivity of models in the active state which is for the contour the starting model; the other problem concerns capturing the characteristic of numerical gradient-based techniques for finding local minima as a result of noise in images and edges which are not concrete. At first, the destination object is segmented roughly using a single image set, one slice at a time, by applying a 2D procedure. In the second step the initial surface is segmented in the beginning. A finite difference method which results in the generation of a simultaneous GA optimizer is used to solve the equation which controls the time change for the contour. In the later stage the simultaneous GA is used to finesse the surface which generates the ultimate result. In order to construct the image, the object surface data and a GA are used where the discontinuous surface is represented geometrically using Fourier descriptors.

In [33] one more technique is presented which also depends upon multiple fitness functions. A co-evolutionary GA uses a multiple energy term based on the fitness function and has two different components. The first component determines its own fitness by making an active comparison of a single universe of chromosomes among themselves; the second one determines the fitness of the whole system. The fitness of the chromosome is obtained as a component of the final solution. While the first component uses only gradient energy, the latter one depends upon three features given below:

1. The distance of the control points of the current chromosome.
2. The best chromosome obtained by using the Euclidean distance function; a term which is used for the minimization of the distance of the subset of end points of the contour in focus now and that of the best one in the universe of populations.
3. A count function that observes the cardinality of the set of Fourier descriptors of the contour which is the result of the crossover operation being performed upon the current and best chromosome from other populations.

However, the evolutionary scheme implemented in [38] is different. Here, the approach involves a multiple stage minimization process of a contour which is active at present. A formulation using quintuples of the ith state transition refers to the contour state. The population of different states is arbitrarily generated in which a GA is run. A contour is extended to a higher state if the latter has less energy than the current state. In all the proposals using swarm intelligence every swarm which is related to some control point is managed by PSO. The local energy of a point in a swarm refers to the cost of that particle. The scanning of control points is performed on a repetitive basis by allocating a search window and then optimizing the local energy. When it is found that no change occurs in all the swarms, the process is terminated. The groups of approaches which use fitness functions differ very much from those which learn energy term weights using metaheuristics. Mostly, the basis for these approaches is the segmentation error derived from the running contours for which parameter adjustment is done without any other intervention. Previously segmented reference images are used to achieve this aim. In [36] this procedure is followed. The use of a segmentation error rate is replaced with a fitness function derived from the distance of the center of gravity from the previously segmented curve. The fitness is considered as the energy of the active contour in many of the evolutionary based techniques. In [38] a four-control-point-based unique active contour is used for which weights are generated through GA. Following the survival of the fittest principle the contour deformation is carried out on the best individual. The fitness function used is given by

$$\frac{1}{(1 + \alpha E_{continuity} + \beta E_{curvature} + \gamma E_{image} + K E_{pressure})} \tag{1}$$

Segmentation tasks determine the fitness function in the approaches using metaheuristics for setting the initial values of the contour models. A Canny edge detector is used in the selection of the curve at the beginning in [20] for optimizing different parameters and the threshold. A fitness function is directly proportional to the product of the weight and the ratio of the maximum value of the length to that of the total length. The approach in [32] is used for the snake initialization method. This in turn sets the number of snakes and the number of control points required. The state of equilibrium of a swarm is gauged by the activity of the swarm. The equilibrium of a swarm is judged and checked as to whether it has dropped below a certain threshold value or not; then the agents are ready for shaping up contours. The adjacent agents to an agent are connected pairwise in some predetermined order to achieve this. The selection of neighbors is done by finding two neighbors on this basis and on factors like distance and angle. The selection of agents is based on distance and so the agents which are closer are preferred. However, as far as angle is concerned, the agents forming wider connections are given preference. The user has to prefix threshold values to put a constraint on the distance and the angle in advance.

3   Segmentation of Histological Images Using a Metaheuristic-Based Level Set Approach

In this section we discuss the segmentation of the hippocampus in histological images given by Mesejo et al., where they follow the idea presented by Ghosh and Mitchell [14]. Here they introduce a new operator called a real-value crossover operator which closely matches the behavior of our chromosomes. The attractive feature of this method is that they use textual features, which is different from the earlier approaches. [28] applied segmentation of the hippocampus of histological images in bi-steps. They used a gray level co-occurrence matrix (GLCM) which produces better results than existing work. Since the size of the histological images is huge, the application of segmentation of text is only applied for a fixed number of repeated steps. Since the structure orientation increases the temporal as well as computational cost during optimization, we do not consider it here. The two main phases in this method are training and proper segmentation. The computation of the average shape is done on the basis of the physically obtained segments during the process of training. The computation of a suitable hippocampus is carried out and from it the principal modes of variation and median texture are obtained. Principle component analysis (PCA) is used on a group of signed representations of distance to generate a prototype of a segmenting curve. The weights to combine the mean shape and shape variables into single unidirectional vectors in the process of segmentation are obtained by using metaheuristics. Also, a comparison is carried out between the texture enclosed by the evolving contour and the representative texture.

3.1 Training Phase

3.1.1 Shape

At first, from the training set we have to extract the shape. For this we have to represent the contours in the zero level set of the signed distance function $P_i(u, v)$ where $1 \leq$ and $\leq v$ are the pixel coordinates and n denotes the cardinality of the set of training contours required to obtain shape variability. The signum function is used for representing the shape, in which shape boundaries are linked to the zero level set of a signum function. The insides and outside of the object are assigned negative and positive distances respectively. The mean level set function is defined as:

$$Q_i(x, y) = \frac{1}{n} \sum_{i=1}^{N} P_i(u, v) \tag{2}$$

Mean offset functions are computed from the relation (3) as follows:

$$\tilde{Q}_i = Q_i - \bar{P} \tag{3}$$

We elaborate the above construction through an example. Suppose M is the image size, which is say $M = M_1 \times M_2$. Then we have to resize the image to say 500×500 pixels. To form a column vector b_i of size $1 \times M$ the columns of M are put in a stack one after another in their order of occurrence. The shape variability matrix V (of size $M \times m$) is derived from these column vectors as

$$V = [b_1, b_2, b_3 \ldots, b_m]$$

Shape variance is given by Eq. (4) and it looks like an eigenvalue decomposition on V:

$$\frac{1}{m} V V^T = U P U^T \tag{4}$$

where U is an $M \times m$ matrix having m orthogonal modes of shape variation as its columns, P, whose elements are eigenvalues in an $m \times m$ diagonal matrix and the columns of U are the corresponding eigenvectors. In [24] a smaller matrix is considered for the computation of the eigenvectors; the eigenshapes are given by (5).

$$\{a_1, a_2, a_3 \ldots, a_n\} \tag{5}$$

In [41] the further computation of the level function representing the segmenting curve is derived as in (6)

$$a[w] = \bar{a} + \sum_{i=1}^{k} w_j a_j \qquad (6)$$

It is easy to see that now the role of the metaheuristic is confined to obtaining the values of w which put the value of a fitness function to be defined in the test phase.

3.1.2 Texture

Visual points when arranged in some pattern with several replica or normal patterns are said to form a texture. GLCM is a procedure which uses statistical techniques to check the characteristics of a texture by taking into account the 3D relationship of the constituents. GLCM was introduced by Haralick et al. [17] is a method that uses the features describing a texture and are comprised of feature values. The Co-occurance Matrix (CM) represents the spatial relationship between two gray levels. The (m, n)th entry in a GLCM represents the cardinality of occurrences of the m and n pair of gray levels which are separated from each other by not more than a predefined distance in a specific direction.

Let us consider an example again. We follow the textural priors narrated in [28], which shows high end performance. Eleven textual features which are used to encode the training pattern as a vector are as follows. These features can be broadly categorized as first-order and second-order measures. Under the first category we have the features of standard deviation, skewness, kurtosis, entropy, coefficient of variation, and energy; whereas the features which come under the second category are contrast, correlation, energy, and homogeneity from GLCM. For example, using $(1, 1)$ as the spatial relationship (i.e. $\theta = 315°$ and $d = 1$ pixel), where a window of dimension 30×30 pixels is used. The focus was on capturing the textural essence of the hippocampus in the course of segmenting and enclosing the texture by the DM surface with the ideal texture of the training set. To realize this median texture an image density of p is taken in the training set with an arbitrary selection procedure inside the hippocampus; a sample of size t were selected through them as pixels-of-interest. The dimension of the generated matrix is $(a \cdot b) \times c$, where c is the cardinality of the training image set, a is the cardinality of the textual feature set, and b is that of the selected point set. Specifically in the example, a is chosen to be 11 and b is taken as 100. The value of the median computed from this matrix represents the texture in the hippocampus with the assumption that points having values nearer to it are considered as being in the hippocampus.

3.1.3 Test Phase

This phase is related to segmenting the object in focus in a justified manner. The outer boundary of the hippocampus is used to fit a hidden model of the contour being guided by a metaheuristic. In order to achieve this weights are generated through a combination of mean and variability. A combination of region and texture based terms are used to generate the fitness function. For the first one the model in [9] is used, whereas for the second one a standard distance function between the textures related to our contour and the median texture found in the training set is used. The Euclidean distance is used to measure the median texture extracted from the training set (T) and the actual texture enclosed by the evolving contour (t(C)). Using this functional, our model takes intensity and texture criteria simultaneously into account. In this case, both terms have been weighted equally.

At the end, a quick step is used for refinement, which requires 50 repeated applications of the local procedure introduced in Chan and Vese [43]. This step considers only the points in the nearer area of the boundary, and extradition connected components is done for components having an area smaller than a pre-assigned value (in terms of pixels).

4 Conclusion

DMs are procedures considered for segmenting, which uses a curve aimed at optimizing positively the overlapping with the actual contour of an object under consideration inside an image. This work comprised of dealing with metaheuristic applications in segmenting images using live contour models. Our presentation also includes the segmentation of histological images using a metaheuristic based level set approach.

References

1. C. Aguilera, R. Sanchez, E. Baradit, Detection of knots using x-ray tomographies and deformable contours with simulated annealing. Wood Res **53**(2), 57–66 (2008)
2. T. Bäck, H.-P. Schwefel, An overview of evolutionary algorithms for parameter optimization. Evol. Comput. **1**(1), 1–23 (1993)
3. L. Ballerini, Genetic snakes for medical images segmentation, in *Mathematical Modeling and Estimation Techniques in Computer Vision, SPIE*, vol. 3457 (1998), pp. 284–295
4. L. Ballerini, An automatic system for the analysis of vascular lesions in retinal images, in *Nuclear Science Symposium, 1999. Conference Record. 1999 IEEE*, vol. 3 (IEEE, Piscataway, 1999), pp. 1598–1602
5. L. Ballerini, Genetic snakes for color images segmentation, in *Applications of Evolutionary Computing* (Springer, Berlin, 2001), pp. 268–277
6. N. Barreira, M.G. Penedo, Topological active volumes. EURASIP J. Adv. Signal Process. **2005**(13), 1937–1947 (2005)

7. J. Bredno, T.M. Lehmann, K. Spitzer, Automatic parameter setting for balloon models, in *Proceedings of SPIE*, vol. 3979 (2000), pp. 1185–1194
8. S. Cagnoni, A. Dobrzeniecki, R. Poli, J. Yanch, Genetic algorithm-based interactive segmentation of 3d medical images. Image Vis. Comput. **17**(12), 881–895 (1999)
9. T.F. Chan, L.A. Vese, Active contours without edges. IEEE Trans. Image Process. **10**(2), 266–277 (2001)
10. D.-H. Chen, Y.-N. Sun, A self-learning segmentation framework–the Taguchi approach. Comput. Med. Imaging Graph. **24**(5), 283–296 (2000)
11. T.R. Crimmins, A complete set of Fourier descriptors for two-dimensional shapes. IEEE Trans. Syst. Man Cybern. **12**(6), 848–855 (1982)
12. Z. Delu, Z. Zhiheng, X. Shengli, Vector bundle constraint for particle swarm optimization and its application to active contour modeling. Prog. Nat. Sci. Mater. Int. **17**(10), 1220–1225 (2007)
13. L.J. Fogel, A.J. Owens, M.J. Walsh, *Artificial Intelligence Through Simulated Evolution* (John Wiley & Sons, Hoboken, 1966)
14. P. Ghosh, M. Mitchell, Segmentation of medical images using a genetic algorithm, in *Proceedings of the 8th annual conference on Genetic and evolutionary computation* (ACM, New York, 2006), pp. 1171–1178
15. F. Glover, M. Laguna, R. Martí, Scatter search, in *Advances in Evolutionary Computing* (Springer, Berlin, 2003), pp. 519–537
16. D.E. Goldberg, *Genetic Algorithms in Search, Optimization, and Machine Learning* (Addison-Wesley, Reading, 1989)
17. R.M. Haralick, K. Shanmugam, et al., Textural features for image classification. IEEE Trans. Syst. Man Cybern. **SMC-3**(6), 610–621 (1973)
18. F. Herrera, M. Lozano, J.L. Verdegay, Tackling real-coded genetic algorithms: operators and tools for behavioural analysis. Artif. Intell. Rev. **12**(4), 265–319 (1998)
19. J.H. Holland, *Adaptation in Natural and Artificial Systems: An Introductory Analysis with Applications to Biology, Control, and Artificial Intelligence* (MIT press, Cambridge, 1992)
20. C.-Y. Hsu, C.-Y. Liu, C.-M. Chen, Automatic segmentation of liver pet images. Comput. Med. Imaging Graph. **32**(7), 601–610 (2008)
21. A.R. Hussain, Optic nerve head segmentation using genetic active contours, in *International Conference on Computer and Communication Engineering, 2008. ICCCE 2008* (IEEE, Piscataway, 2008), pp. 783–787
22. J. Kennedy, R. Eberhart, Particle swarm optimization, in *IEEE International Conference on Neural Networks*, vol. 4 (1995), pp. 1942–1948
23. R.A. Kirsch, Computer determination of the constituent structure of biological images. Comput. Biomed. Res. **4**(3), 315–328 (1971)
24. M.E. Leventon, Statistical models in medical image analysis, PhD thesis, Massachusetts Institute of Technology, 2000
25. P. Liatsis, C. Ooi, Tracking moving objects with co-evolutionary snakes, in *Video/Image Processing and Multimedia Communications 4th EURASIP-IEEE Region 8 International Symposium on VIPromCom* (IEEE, Piscataway, 2002), pp. 325–332
26. L.A. MacEachern, T. Manku, Genetic algorithms for active contour optimization, in *Proceedings of the 1998 IEEE International Symposium on Circuits and Systems, 1998. ISCAS'98*, vol. 4 (IEEE, Piscataway, 1998), pp. 229–232
27. U. Maulik, Medical image segmentation using genetic algorithms. IEEE Trans. Inf. Technol. Biomed. **13**(2), 166–173 (2009)
28. P. Mesejo, R. Ugolotti, S. Cagnoni, F. Di Cunto, M. Giacobini, Automatic segmentation of hippocampus in histological images of mouse brains using deformable models and random forest, in *25th International Symposium on Computer-Based Medical Systems (CBMS), 2012* (IEEE, Piscataway, 2012), pp. 1–4
29. E. Mezura-Montes, C.A.C. Coello, Constraint-handling in nature-inspired numerical optimization: past, present and future. Swarm Evol. Comput. **1**(4), 173–194 (2011)
30. Z. Michalewicz, *Genetic Algorithms + Data Structures = Evolution Programs* (Springer Science & Business Media, Berlin, 2013)

31. P. Moscato, C. Cotta, A gentle introduction to memetic algorithms, in *Handbook of Meta-heuristics* (Springer, Berlin, 2003), pp. 105–144
32. K.-J. Mun, H.T. Kang, H.-S. Lee, Y.-S. Yoon, C.-M. Lee, J.H. Park, Active contour model based object contour detection using genetic algorithm with wavelet based image preprocessing. Int. J. Control. Autom. Syst. **2**(1), 100–106 (2004)
33. C. Ooi, P. Liatsis, Co-evolutionary-based active contour models in tracking of moving obstacles, in *Proceedings of International Conference on Advanced Driver Assistance Systems* (2001), pp. 58–62
34. J. Paredis, Coevolutionary computation. Artif. Life **2**(4), 355–375 (1995)
35. M.E. Plissiti, D.I. Fotiadis, L.K. Michalis, G.E. Bozios, An automated method for lumen and media-adventitia border detection in a sequence of ivus frames. IEEE Trans. Inf. Technol. Biomed. **8**(2), 131–141 (2004)
36. J.-J. Rousselle, N. Vincent, N. Verbeke, Genetic algorithm to set active contour, in *Computer Analysis of Images and Patterns* (Springer, Berlin, 2003), pp. 345–352
37. R. Storn, K. Price, Differential evolution–a simple and efficient adaptive scheme for global optimization over continuous spaces: technical report tr-95-012. International Computer Science, Berkeley, California, 1995
38. M. Talebi, A. Ayatollahi, Genetic snake for medical ultrasound image segmentation. *Image Analysis and Recognition* (Springer, Berlin, 2011), pp. 48–58
39. L. Tang, K. Wang, G. Feng, Y. Li, An image segmentation algorithm based on the simulated annealing and improved snake model, in *International Conference on Mechatronics and Automation, 2007. ICMA 2007* (IEEE, Piscataway, 2007), pp. 3876–3881
40. G.M. Teixeira, I.R. Pommeranzembaum, B.L. de Oliveira, M. Lobosco, R.W. Dos Santos, Automatic segmentation of cardiac MRI using snakes and genetic algorithms, in *International Conference on Computational Science* (Springer, Berlin, 2008), pp. 168–177
41. A. Tsai, A. Yezzi, W. Wells, C. Tempany, D. Tucker, A. Fan, W.E. Grimson, A. Willsky, A shape-based approach to the segmentation of medical imagery using level sets. IEEE Trans. Med. Imaging **22**(2), 137–154 (2003)
42. M. Vera, A. Bravo, R. Medina, Myocardial border detection from ventriculograms using support vector machines and real-coded genetic algorithms. Comput. Biol. Med. **40**(4), 446–455 (2010)
43. X.-F. Wang, D.-S. Huang, H. Xu, An efficient local Chan–Vese model for image segmentation. Pattern Recogn. **43**(3), 603–618 (2010)
44. J.-M. Yang, C.-Y. Kao, Integrating adaptive mutations and family competition into genetic algorithms as function optimizer. Soft Comput. A Fusion Found. Methodol. Appl. **4**(2), 89–102 (2000)
45. V. Zografos, Comparison of optimisation algorithms for deformable template matching. *Advances in Visual Computing* (Springer, Berlin, 2009), pp. 1097–1108

Hybridization of the Univariate Marginal Distribution Algorithm with Simulated Annealing for Parametric Parabola Detection

S. Ivvan Valdez, Susana Espinoza-Perez, Fernando Cervantes-Sanchez, and Ivan Cruz-Aceves

Abstract This chapter presents a new hybrid optimization method based on the univariate marginal distribution algorithm for a continuous domain, and the heuristic of simulated annealing for the parabola detection problem. The hybrid proposed method is applied to the DRIVE database of retinal fundus images to approximate the retinal vessels as a parabolic shape. The hybrid method is applied separately using two different objective functions. Firstly, the objective function only considers the superposition of pixels between the target pixels in the input image and the virtual parabola; secondly, the objective function implements a weighted restriction on the pixels close to the parabola vertex. Both objective functions in the hybrid method obtain suitable results to approximate a parabolic form on the retinal vessels present in the retinal images. The experiments show that the parabola detection results obtained from the proposed method are more robust than those obtained by the comparative method. Additionally, the average execution time achieved by the proposed hybrid method (1.57 s) is lower than the computational time obtained by the comparative method on the database of 20 retinal images, which is of interest to computer-aided diagnosis in clinical practice.

Keywords Estimation of distribution algorithms · Hough transform · Hybrid optimization · Image analysis · Parabola detection · Simulated annealing

S. I. Valdez
Universidad de Guanajuato, División de Ingenierías, Salamanca, Guanajuato, Mexico

S. Espinoza-Perez
Universidad del Papaloapan, Ingeniería en Computación, Loma Bonita, Oaxaca, Mexico

F. Cervantes-Sanchez
Centro de Investigación en Matemáticas (CIMAT), Valenciana, Guanajuato, Mexico

I. Cruz-Aceves (✉)
CONACYT, Centro de Investigación en Matemáticas (CIMAT), Valenciana, Guanajuato, Mexico
e-mail: ivan.cruz@cimat.mx

© Springer International Publishing AG, part of Springer Nature 2018
S. Bhattacharyya (ed.), *Hybrid Metaheuristics for Image Analysis*,
https://doi.org/10.1007/978-3-319-77625-5_7

1   Introduction

The automatic detection of parametric objects is an essential task in different research areas such as medical and natural image analysis. In general, this problem can be addressed in two steps: the definition of the parametric equation of the target object (which is not an easy task), and the definition of a search strategy. In the first step, the boundaries of the target object have to be clearly detected in order to perform the matching process with the parametric equation. This task is commonly performed by using edge detection techniques such as Sobel or Canny operators. To perform the second step, an exhaustive or heuristic strategy can be applied.

In the literature, the Hough transform (HT) is the most highlighted method for parametric object detection [1–3]. The HT is a standard technique for parametric shape recognition which is useful for image analysis and computer vision. HT was first applied to straight line detection [4], and later used to detect circles [5–7], ellipses [8], and parabolas [9, 10]. The main advantages of the HT method are that it is insensitive to noise and it is easy to implement; however, the main disadvantage is the execution time, since the computational complexity is $\mathscr{O}(n^s)$, where s is the number of unknown variables in the parametric equation defining the target object.

An important application of the HT for parabola detection is the approximation of the shape of blood vessels in fundus images of the retina. It is known that some pathologies, such as diabetic retinopathy, affect the shape of the vessels in the retina and because of this the shape itself can be used to aid the diagnosis of those diseases [9, 10]. In medical practice, the approximation of the shape of blood vessels in retinal images is an exhaustive manual task which requires the visual detection of representative features on the retinal vasculature. This task can be automated using the HT transform embedded within a computerized aided system in order to reduce the time and intensive labor taken by a specialist during the diagnosis of diseases.

On the other hand, to solve the drawback of computational complexity of the HT method, several strategies have been introduced involving least squares [11], the pruning-and-voting strategy [12], randomized techniques [13–16], and population methods based on particle swarm optimization [17], electromagnetism optimization [18], genetic algorithms [19], an artificial immune system [20], differential evolution [21], and estimation of distribution algorithms (EDAs) [22, 23].

In general, the performance of the population-based methods is very suitable in terms of computational time and detection accuracy, since they are useful for avoiding the local minima problem. In the present chapter, a new hybrid optimization method based on an estimation of the distribution algorithm for a continuous domain; simulated annealing (SA) is introduced to detect parabolic shapes on retinal fundus images. Since the proposed method is a stochastic optimization technique, the introduced fitness function is based on the shape maximization between a virtual parabola and the target object of the input image. In addition, the proposed method is compared with different state-of-the-art parametric detection methods in terms of computational time, robustness, and detection accuracy.

The chapter is organized as follows. In Sect. 2, the parabola detection problem is introduced along with the techniques of SA and estimation of distribution algorithms, which are explained in detail. In Sect. 3, the proposed hybrid method for parabola detection is introduced and analyzed. Computational results are presented and discussed in Sect. 4, and conclusions are given in Sect. 5.

2 Background

This section introduces the fundamentals of the parabola detection problem and two state-of-the-art estimation of distribution algorithms along with the local search method known as SA.

2.1 Parabola Detection Problem

The parabola detection problem can be defined as a matching problem. Hence, we draw a parabola in a binary image, then we compute the number of white pixels which match with the white pixels in the target image. In order to draw the parabola we require a set of four parameters $\mathbf{x} = [x_1 = a, x_2 = b, x_3 = c, x_4 = \theta]$ used as shown in Eq. (1).

$$
\begin{aligned}
\hat{y} &= a\hat{x}^2 + b\hat{x} + c, \\
x_v &= -\frac{b}{2a}, \\
y_v &= ax_v^2 + bx_v + c, \\
x &= \text{integer}(\cos(\theta)(\hat{x} - x_v) - \sin(\theta)(\hat{y} - y_v) + x_v), \\
y &= \text{integer}(\sin(\theta)(\hat{x} - x_v) + \cos(\theta)(\hat{y} - y_v) + y_v),
\end{aligned}
\tag{1}
$$

where $[a, b, c]$ are used to define a parabola aligned to the y-axis, which is then rotated at a θ angle, with respect to the vertex (x_v, y_v). Consider a target image with height n_{row} and width n_{col}. We draw a parabola with a set of coordinates $[x_i, y_i]$ for $i = 1, 2, \ldots, n_{\text{coord}}$, where $x_i \in \{1 \ldots n_{\text{row}}\}$ and $y_i \in \{1 \ldots n_{\text{col}}\}$. Then, we count the number of pixels p that are different from 0 in the parabola and the target images according to Eq. (2). This equation is the *objective function*; notice that $[x_i, y_i]$ depend on $\mathbf{x} = [x_1 = a, x_2 = b, x_3 = c, x_4 = \theta]$.

$$
f(x_i, y_i) = \sum_{i=1}^{n_{\text{coord}}} p_{\text{par}}(x_i, y_i) p_{\text{target}}(x_i, y_i),
\tag{2}
$$

where $p_{\text{par}}(x_i, y_i)$ and $p_{\text{target}}(x_i, y_i)$ are pixel values in the parabola and target image, respectively.

2.1.1 Hough Transform

In image analysis, the HT is the most commonly used strategy to detect parametric forms such as lines or circles [1]. On the other hand, the HT can also be extended for detecting parabolic shapes. In general, HT requires the parametric equation of the object to be detected in the Cartesian or polar coordinate system. For instance, to detect lines, circles, or parabolas in Cartesian coordinates, the following equations can be evaluated, respectively:

$$y = mx + b \tag{3}$$

$$r^2 = (x - a)^2 + (y - b)^2 \tag{4}$$

$$(y - y_0)^2 = 4a(x - x_0) \tag{5}$$

In order to detect parametric objects in images by using the HT algorithm, the input image must be binary. The pixels with intensity different to zero represent the potential object to be detected, while all the pixels with intensity zero represent the background image, which are irrelevant for the process.

The resolution of the search space plays an important role, since it is used for testing all possible parameter combinations (exhaustive search strategy). Consequently, there is a trade-off between detection precision and computational time. High precision obtains good detection results involving high execution time, but low precision may not find the target object.

The generic procedure to perform the HT can be described as follows:

1. Determine the equation of the parametric object to be evaluated and the number of unknown parameters n.
2. Initialize the dimension of the accumulator array according to n.
3. Compute for each pixel (x, y) of interest the parametric equation.
4. Increment the value of the accumulator array for the set of unknown parameters.
5. Find the sets of parameters with the highest values.
6. Determine the best set of parameters for the parametric object (typically by applying a threshold or a local maxima strategy).

The HT strategy presents two main disadvantages. Firstly, the high computation time of the exhaustive search, and secondly, the method to determine the optimal set of values, where the most widely used strategy to find it is the local maxima method.

2.2 Simulated Annealing

SA is a widely known algorithm first proposed for discrete domains by Kirkpatrick et al. [24]; other researchers have presented different modifications and versions. One of the best performed in continuous domains is the one proposed by Corana et al. [25]. Nevertheless, SA is a global optimizer for multimodal functions; we use it as a local optimizer, considering that it searches in the vicinity of the best solution at each iteration of the optimization process. SA is shown in Algorithm 1.

The initial temperature is set as: $T_0 = -\hat{f}/(2\log(0.2))$ where $\hat{f}$ is the average of the objective function value of the current population. Notice that the initial temperature changes in every iteration. The step size is computed as suggested by Corana et al. [25]. Finally, the well known Metropolis criterion is as follows:

Algorithm 1: Simulated annealing

Input: $T_0 = $ A high temperature.
$x^{inf}, x^{sup} = $ Inferior and superior limits respectively.

1   $t = 0$;
2   Initial solution $\mathbf{x}_{best} = \mathbf{x}_0 = [x_1, x_2, \ldots, x_n]$;
3   Evaluate and initialize the best objective function value $f_{best} = f(\mathbf{x}_t)$;
4   $\hat{\mathbf{x}} = \mathbf{x}_t$;
5   **while** *Stopping criterion is not met* **do**
6   **for** $T = 1 \ldots N_T$ **do**
7   **for** $n_s = 1..N_s$ **do**
8   **for** $i = 1..n$ **do**
9           $e_i \sim U(-v_i, v_i)$;
10   **while** $(\hat{x}_i + e_i) < x_i^{inf} \vee (\hat{x}_i + e_i) > x_i^{sup}$ **do**
11             $e_i \sim U(x^{inf}, x^{sup})$;
12   Evaluate $\hat{\mathbf{x}}$;
13   **if** $f(\hat{\mathbf{x}}) < f_{best}$ **then**
14             $f_{best} = f(\hat{\mathbf{x}})$;
15             $\mathbf{x}_{best} = f(\hat{\mathbf{x}})$;
16   Accept the solution $\hat{\mathbf{x}}$ according to the Metropolis criterion;
17   Adjust the step size v;
18     $T = \alpha_T T$;
19   Update $\mathbf{x}_t = \mathbf{x}_{best}$ and $f(\mathbf{x}_t) = f_{best}$;

---

1. Let $\mathbf{x}_t$ be the current solution and $\hat{\mathbf{x}}$ the perturbed solution.
2. $\Delta f = f(\mathbf{x}_t) - f(\hat{\mathbf{x}})$.
3. **If** $(\Delta f(\mathbf{x}) \leq 0)$ **then** $\mathbf{x}_{t+1} = \hat{\mathbf{x}}$.
4. **else** The perturbed solution is accepted with probability $\exp\left(\frac{-\Delta f}{T}\right)$, that is to say the perturbed solution replaces the current one.

The Metropolis criterion permits us to accept wrong solutions rather than the current one with the aim of avoiding local minima problems. The algorithm stops if the objective function is not improved more than $\epsilon = 1$ in five consecutive temperature reductions or reaches 30 function evaluations. The SA algorithm is presented in the Appendix in the C programming language.

2.3 Estimation of Distribution Algorithms

Estimation of distribution algorithms (EDAs) are population-based methods used to solve optimization problems. The main difference with evolutionary computation techniques is the fact that the crossover and mutation operators are not required, since the new potential solutions are generated by building probabilistic models based on the statistical information of promising solutions [26–28]. In this work, we focus on the univariate marginal distribution algorithm (UMDA) in both discrete and continuous domains.

2.3.1 Univariate Marginal Distribution Algorithm

The UMDA is a search strategy derived from EDAs to solve linear optimization problems with not many significant dependencies [29]. This stochastic method is a population-based strategy where a marginal probability is computed at each generation [30]. The probabilistic model is calculated from a subset of individuals (potential solutions) in order to generate the next population of the iterative process. Similar to evolutionary computation techniques, UMDA uses a fitness function and binary encoding to represent the individuals. Consequently, the genes of the individuals are randomly initialized between {0, 1} with a uniform probability.

The selection step is applied to select a subset of individuals. This operator uses the truncation strategy, which orders the solutions according to their fitness. Subsequently, the estimation of the univariate marginal probabilities P are computed using the subset of individuals. The marginal probability model for independent variables can be defined as follows:

$$P(x) = \prod_{i=1}^{n} P(X_i = x_i), \qquad (6)$$

where $x = (x_1, x_2, \ldots, x_n)^T$ is the binary value of the ith bit in the solution, and X_i is the ith uniform random value of the vector X.

Finally, UMDA generates a new population from the estimated marginal probability model. The process is iteratively performed until a convergence criterion is satisfied, and the best individual (elite) is the one with the best fitness along generations.

According to the above description, UMDA can be implemented as follows:

1. Initialize number of individuals n.
2. Initialize number of generations t.
3. Initialize selection rate $[0, 1]$.
4. Initialize the individuals into the predefined search space.
5. Select a subset of individuals S of $m \leq n$ according to the selection rate.
6. Compute the marginal probabilities $p_i^s(x_i, t)$ of S.
7. Generate n new individuals by computing $p(x, t + 1) = \prod_{i=1}^{n} p_i^s(x_i, t)$.
8. Stop if convergence criterion is satisfied, otherwise repeat steps (4)–(7).

2.3.2 Univariate Marginal Distribution Algorithm for Continuous Domains (UMDA$_c$)

The univariate marginal distribution algorithm for continuous domains (UMDA$_c$) is an evolutionary algorithm from the family of EDAs. It was proposed by Larrañaga and Lozano [26]. In this case, we use a particular version which a priori defines the normal probability function; this version is named UMDA$_{Gc}$. The UMDA$_{Gc}$ considers that variables are independent of each other; the joint probability function can then be written as:

$$f_{\mathcal{N}}(x, \theta) = \prod_{i=1}^{n} \frac{1}{\sqrt{2\pi}\sigma_i} e^{-\frac{1}{2}\left(\frac{x_i - \mu_i}{\sigma_i}\right)^2} \tag{7}$$

Thus, the parameters μ_i and σ_i must be estimated from the selected set for each dimension i. The UMDA$_{Gc}$ is shown in Algorithm 2. At line 2, the initial population is sampled from a uniform distribution defined inside given limits. At line 3, the population is evaluated, then half of the best solutions are selected in the S set, and the elite individual, x_{best}, f_{best}, is stored. Then, for n_{gen} generations, the vectors of means μ and standard deviations σ are estimated via maximum likelihood estimators over S. At line 7, the new population is formed by the combination of $n_{\text{pop}} - 1$ sampled candidate solutions and the elite individual, and so on. Figure 1 illustrates the flow of Algorithm 2 for the UMDA$_{Gc}$ strategy.

Algorithm 2: The univariate marginal distribution for continuous domains using a normal distribution

---

Data: D=Number of dimensions.
       n_{pop}=Population size.
       n_{gen}=Number of generations.
       x_{inf} =Inferior search limits.
       x_{sup} =Superior search limits.
Result: x_{best} and f_{best}. Optimum approximation and its objective function value.

1 $t = 0$;
2 $X^t \sim U(x_{inf}, x_{sup})$;
3 $F^t = Evaluate(X^t)$;
4 $[S^t, x_{best}^t, f_{best}^t] = Selection(X^t, F^t)$;
5 **for** $1..n_{gen}$ **do**
6     $[\mu^t, \sigma^t] = Parameter\,Estimation(S^t)$;
7     $X^{t+1} = [x_{best}^t, Sampling(\mu^t, \sigma^t, n_{pop} - 1)]$;
8     $t = t + 1$;
9     $F^t = Evaluate(X^t)$;
10    $[S, x_{best}^t, f_{best}^t] = Selection(X^t, F^t)$;

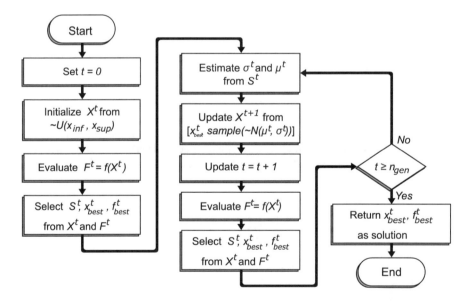

Fig. 1 Flowchart of the algorithm of the univariate marginal distribution algorithm for continuous domains using a normal distribution

2.3.3 Applications

The EDAs have proven to be very effective for solving high-dimensional optimization problems. In the literature, different mathematical functions have been introduced for testing optimization algorithms. In this section, to illustrate the

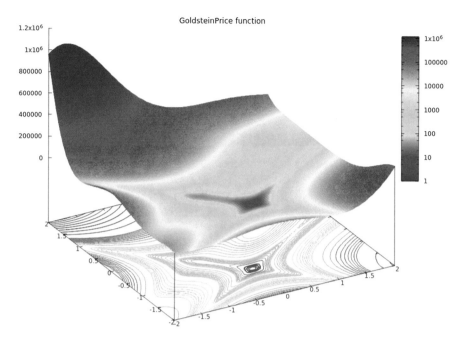

Fig. 2 Goldstein–Price function in two dimensions including level plot in the range $[-2, 2]$ for each independent variable

implementation of UMDA_{Gc} in an optimization task, the *2D Goldstein–Price* function is introduced.

Commonly, the range for each variable of the test function is $X_1 \in [-2, 2]$, $X_2 \in [-2, 2]$, and the optimal value is located on $f(0, -1) = 3$. The *2D Goldstein–Price* test function is defined below in Eq. (8), and illustrated in Fig. 2.

$$f(x_1, x_2) = \left[1 + (x_1 + x_2 + 1)^2 (19 - 14x_1 + 3x_1^2 - 14x_2 + 6x_1x_2 + 3x_2^2) \right]$$
$$\times \left[30 + (2x_1 - 3x_2)^2 (18 - 32x_1 + 12x_1^2 + 48x_2 - 36x_1x_2 + 27x_2^2) \right].$$

$$(8)$$

Moreover, to solve the Goldstein–Price function, the population of the UMDA_{Gc} strategy is randomly distributed in the search space and also it is encoded using the two dimensions of the problem for each individual. The numerical example of the optimization process using the real-coded UMDA_{Gc} is illustrated in Fig. 3. In the illustration, the ESTIMATION OF THE MARGINAL DISTRIBUTIONS block marks each individual in S^t with a cross on the level curves of the Goldstein–Price function. The convergence to the optimal solution is exemplified as the reduction of the dispersion of the individuals within the search space throughout three generations. The reduction of the dispersion is represented as the narrowing of the distribution curves present in the margin of the block.

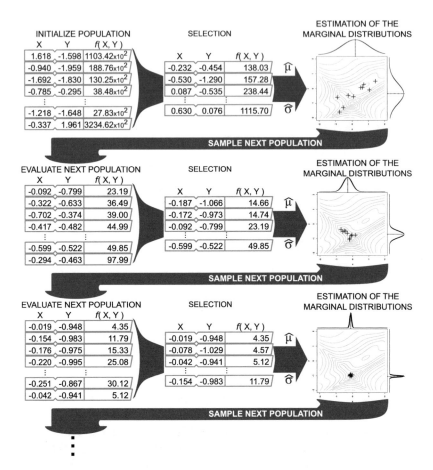

Fig. 3 Numerical example for solving the 2D Goldstein–Price function using UMDA$_{Gc}$ as an optimization strategy

3 Proposed Hybrid Method

In this section, the proposed hybrid method based on the UMDA and SA is presented. In the first stage, the hybrid scheme which is the main contribution of the present work is explained in detail. Finally, the process to adapt the parametric parabola problem to be solved by the proposed method is introduced.

3.1 Hybridization of UMDA$_c$ with SA

The main proposal is a hybridization of UMDA$_{Gc}$ with SA as follows. At each generation the best solution from UMDA$_{Gc}$ (the elite) is used as a starting point for SA.

In the case when the elite individual is improved, this solution is always inserted in the selected set and stored. In addition, for the annealing schedule, the initial temperature is computed by using the objective function values in the last population in $UMDA_{Gc}$. In the same vein, the inferior and superior search limits for all variables of SA are computed by using the standard deviation vectors, computed in the last $UMDA_{Gc}$ iteration, as follows: $x_{inf}^t = x_{best}^t - 0.05sd^t$ and $x_{sup}^t = x_{best}^t + 0.05sd^t$. According to these limits, the search performed by SA occurs in a small vicinity of the current best solution. In addition, the smaller the standard deviation of the current population is, the smaller the vicinity in the SA algorithm is. Thus, both algorithms take advantage of each other, the $UMDA_{Gc}$ is used to provide an adequate temperature, initial point, and search limits of SA, while SA often improves the best solution in the UMDA population, a bias in the search toward the most promising regions.

Algorithm 3 describes the general hybrid proposal. Lines 5 and 12 show where the SA step is; as a result of this step the selected set $\hat{S}^t$ and the elite x^{best} could be updated, if the best solution from $UMDA_{Gc}$ is improved; in such a case this new elite is inserted into the selected set $\hat{S}^t$ and, as a consequence, used to compute the $UMDA_{Gc}$ parameters, as is shown in Line 7.

This hybrid algorithm could be applied to any optimization problem, and it requires as input the population size which is set to 30 in all our reported experiments, a maximum number of generations which is set to 30, and search limits.

For the particular case of using this algorithm for parabola detection, the search limits are computed for each target image as is shown in Line 2; this procedure is detailed in Sect. 3.2.

Algorithm 3: Hybrid $UMDA_{Gc}$/SA

Data: D=Number of dimensions.
  n_{pop}=Population size.
  n_{gen}=Number of generations.
  x_{inf} and x_{sup}=Inferior and superior limits.
Result: x_{best} and f_{best}. Optimum approximation and its objective function value.
1 $t = 0$;
2 $X^t \sim U(x_{inf}, x_{sup})$;
3 $F^t = Evaluate(X^t)$;
4 $[S^t, x_{best}^t, f_{best}^t] = Selection(X^t, F^t, T)$;
5 $[\hat{S}^t, x_{best}^t, f_{best}^t] = SimulatedAnnealing(x_{best}^t, f_{best}^t)$;
6 **for** $1..n_{gen}$ **do**
7 $\quad [\mu^t, \sigma^t] = ParameterEstimation(\hat{S}^t)$;
8 $\quad X^{t+1} = [x_{best}^t, Sampling(\mu^t, \sigma^t, n_{pop} - 1)]$;
9 $\quad t = t + 1$;
10 $\quad F^t = Evaluate(X^t)$;
11 $\quad [S, x_{best}^t, f_{best}^t] = Selection(X^t, F^t)$;
12 $\quad [\hat{S}^t, x_{best}^t, f_{best}^t] = SimulatedAnnealing(x_{best}^t, f_{best}^t, T)$;

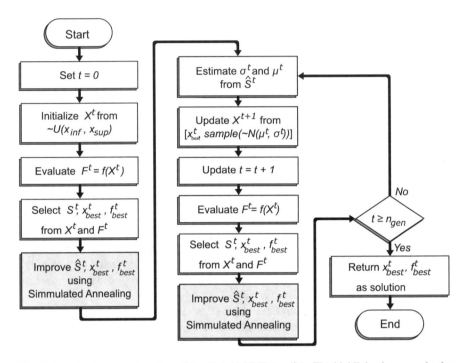

Fig. 4 Flowchart of the algorithm of the Hybrid UMDA$_{Gc}$/SA. The highlighted process in the chart represents the use of SA to improve the best solution found by UMDA$_{Gc}$

The flowchart of the UMDA$_{Gc}$/SA strategy is presented in Fig. 4, where the process of improving the UMDA$_{Gc}$ best solution for each iteration using SA is highlighted. Computationally, the hybridization of the UMDA$_{Gc}$ can expect an average time overhead of up to 48.91% and a mean memory overhead of up to 0.51% due to the improvement of the solution using SA.

3.2 Parametric Parabola Detection Using the Proposed Method

For the purpose of using this proposal for parabola detection, a candidate solution (chromosome) is represented by four variables, as previously mentioned in Sect. 2.1, $\mathbf{x} = [x_1 = a, x_2 = b, x_3 = c, x_4 = \theta]$, and the search limits are necessary.

Algorithm 3 requires the superior and inferior search limits; our proposal computes these limits for each target image automatically; the procedure assumes that the parabola is always vertically aligned, though it could be upwards or downwards; this direction is, also, automatically detected by using the following procedure:

- From the target image we randomly select 10% of the non-zero pixels.
- Using these pixels we compute a least squares parabola fitting, hence we obtain a, b, and c quotients.
- Repeat the two steps above 10 times.
- Then, we get a set of 10 $[a, b, c]$ quotients. We compute the means μ_a, μ_b, and μ_c and the standard deviations σ_a, σ_b, and σ_c. Notice that the sign of the a parameter in the equation $ax^2 + bx + c$ determines whether the parabola opens upwards or downwards. Thus, we define two cases:

  - if $(\mu_a < 0)$

    $a_{\text{inf}} = \mu_a - \sigma_a$,
    $b_{\text{inf}} = \mu_b - \sigma_b$,
    $c_{\text{inf}} = \mu_c - \sigma_c$,
    $a_{\text{sup}} = \mu_a + 4\sigma_a$,
    $b_{\text{sup}} = \mu_b + 4\sigma_b$,
    $c_{\text{sup}} = \mu_c + 4\sigma_c$,
    $x_v = -b_{\text{sup}}/(2a_{\text{sup}})$, notice that this is the x-coordinate of the parabola vertex using the superior limits.
    $y_v = a_{\text{sup}}x_v^2 + b_{\text{sup}}x_v + (\mu_c + 4\sigma_c)$, this is the y-coordinate of the parabola vertex.
    $c_{\text{sup}} = c_{\text{inf}} + (n_{\text{row}} - y_v) - 1$

  - else

    $a_{\text{inf}} = \mu_a - 4\sigma_a$,
    $b_{\text{inf}} = \mu_b - 4\sigma_b$,
    $x_v = -b_{\text{inf}}/(2a_{\text{inf}})$, notice that this is the x-coordinate of the parabola vertex using the inferior limits.
    $y_v = a_{\text{inf}}x_v^2 + b_{\text{inf}}x_v + (\mu_c - 4\sigma_c)$, this is the y-coordinate of the parabola vertex using the inferior limits.
    $c_{\text{inf}} = (\mu_c - 4\sigma_c) - y_v + 1$,
    $a_{\text{sup}} = \mu_a + \sigma_a$,
    $b_{\text{sup}} = \mu_b + \sigma_b$,
    $c_{\text{sup}} = \mu_c + \sigma_c$.

The purpose of this procedure is to find appropriate limits. We know that a least squares fitting is suitable; nevertheless there could be pixels that are not representative of the target image, thus that is the reason for randomly selecting 10% of them with the aim of reducing the bias given by the outliers. Thus, the procedure uses the parameters from 10 parabolas to determine the correct limits using its mean and standard deviation. In addition, by observation we determine that if the parabola opens downwards we must extend the superior limits and vice versa; that is the reason why we set the limit of four standard deviations in one direction and only one in the other.

4   Computational Experiments

In this section, the proposed hybrid method based on the UMDA and SA is applied
on the 20 retinal fundus images of the publicly available DRIVE database [31].
The computational experiments are performed using the Matlab software version
2016, on a computer with an Intel Core i5, 4 GB of RAM, and a 2.4 GHz processor.
To evaluate the proposed method, it is directly compared with the evolutionary
technique proposed by de Jesus Guerrero-Turrubiates et al. [22] based on EDAs in
terms of execution time. Moreover, in order to apply the parabola detection methods,
an automatic vessel segmentation method has to be applied. In the experiments, the
method based on Gaussian matched filters proposed by [28] has been introduced.

On the other hand, the method proposed by de Jesus Guerrero-Turrubiates et
al. [22] uses the UMDA as an optimization strategy to solve the parabola detection
problem. According to this method, the average execution time is 4.5838 s for the set
of retinal images, which was performed by applying the following set of parameters
(Table 1).

Figure 5 illustrates the obtained parabola detection results by applying the
previously mentioned method.

The main contribution of this chapter is the implementation of a hybrid strategy
involving the UMDA method and the heuristic of SA. The UMDA method has been
adopted because of the good results obtained in previous works for the parabola
problem. Here, two different methods are proposed, where the main difference is
the objective function to be evaluated. The detection results of both methods are
presented below.

4.1   Parabola Detection on Retinal Fundus Images

In the first method we use binary images for maximizing the objective function in
Eq. (2). Notice that for binary images any characteristic of the parabola is as good
as any other; that is to say the algorithm looks for matching any segment in the
parabola independent of curvature to the target image. We perform 30 trials for the
set of 20 retinal fundus images. On average, each execution lasts 1.57 s with an
implementation in C language, using the GNU gcc compiler version 6.4. The best
detection results for each set of the trials is reported in Fig. 6.

Table 1 UMDA parameters
for the computational
experiments using the method
proposed in [22]

Parameter	Value
Number of individuals	10
Selection rate	0.6
Maximum number of generations	30

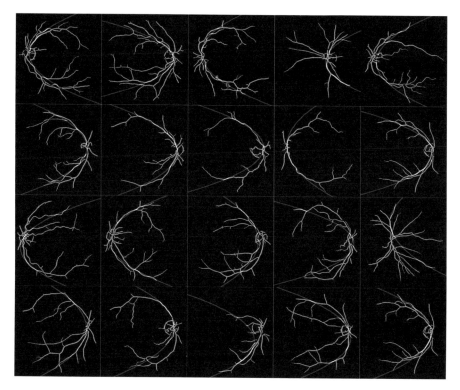

Fig. 5 Parabola detection using the method in [22] on the 20 retinal fundus images of the DRIVE database

4.2 Weighting the Parabola Pixels to Improve Detection

The second proposed method is a manner of weighting pixels of the parabola image, in order to associate a higher importance to the most relevant characteristics. In this sense, it is of interest to find in the target image a similar region to the parabola vertex; we consider that this region, which is close to the change in the derivative sign, is the most important. For this purpose we use Eq. (9); notice that the denominator of the first term is the absolute value of the derivative of the parabola plus 1; we sum to 1 to avoid division by zero. Thus, $\Delta x \in [1, 255]$ is maximum when the derivative is 0; that is to say, when the x_i is the parabola vertex. We use the very same procedure and objective function, but for this case each parabola pixel increases with a different value from the objective function. The results are shown in Fig. 7.

$$\Delta x = 254 \left(\frac{1.0}{|2.0ax_i + b| + 1} + \frac{1}{254} \right). \tag{9}$$

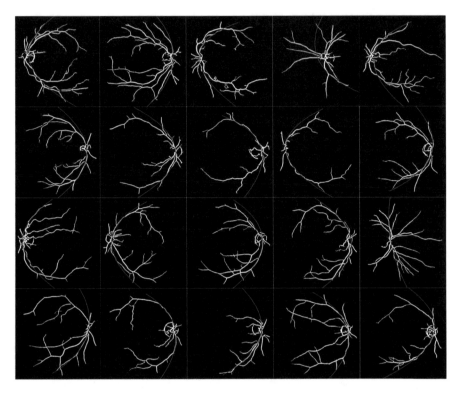

Fig. 6 Parabola detection using the Hybrid UMDA$_{Gc}$/SA on retinal fundus images of the DRIVE database

The quantitative analysis of the two strategies introduced in this work and the method proposed by de Jesus Guerrero-Turrubiates et al. [22] is presented in Table 2. The accuracy metric was defined as the correspondence ratio of vessel pixels superposed by the parametrized parabola. This metric was selected due to the difficulty of defining a parabola ground-truth approved by an expert. According to the comparison analysis, the two hybrid strategies surpass the non-hybrid method. Moreover, the approach of weighting the parabola pixels presents a better performance with the highest correspondence ratio.

Table 3 presents the average execution time of the proposed method and the UMDA [22], Medical Image Processing, Analysis, and Visualization software (MIPAV) [32], and Random Sample Consensus method (RANSAC) [33] methods applied to the parametric parabola detection reported by de Jesus Guerrero-Turrubiates et al. [22]. According to the computational results, the proposed hybrid UMDA$_{Gc}$/SA execution time is lower than the non-hybrid approaches. The highest accuracy and shortest execution time of the hybrid approaches introduced in this work make them more suitable to aid the detection of diseases in fundus images of the retina in medical practice.

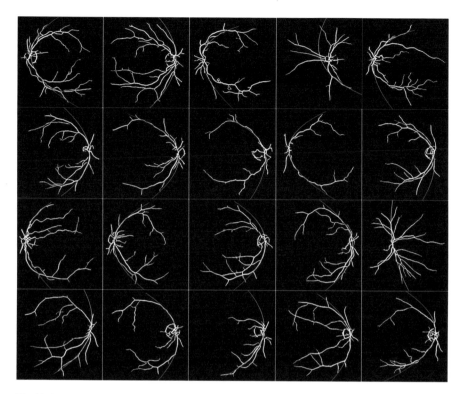

Fig. 7 Parabola detection using the second Hybrid UMDA$_{Gc}$/SA strategy on retinal fundus images of the DRIVE database

Table 2 Comparative analysis of the average accuracy using 20 images of the DRIVE database of retinal fundus images

Method	Average accuracy (%)
Guerrero-Turrubiates method [22]	1.3047
Hybrid UMDA$_{Gc}$/SA	2.4800
Hybrid UMDA$_{Gc}$/SA weighting pixels	**2.5288**

The value in bold represents the best (highest) average accuracy

Table 3 Comparative analysis of the average execution time using the DRIVE database of retinal fundus images

Method	Average execution time (s)
Proposed method	**1.57**
Guerrero-Turrubiates method [22]	4.5838
MIPAV [32]	43.65
RANSAC [33]	16.66

The value in bold represents the best (lowest) average execution time

5 Concluding Remarks

In this chapter a new hybrid optimization method based on the UMDA for continuous domains and SA has been proposed for the parabola detection problem. The proposed method was applied to the DRIVE database of retinal fundus images, using two different objective functions: firstly, only taking into account the superposition of pixels, and secondly, by weighting the detection of the parabola vertex. Both strategies yield suitable results in the approximation of the parabolic shape of the retinal images, obtaining more robust results than the comparative method. In addition, the average computational time (1.57 s) of the proposed method outperforms that of the comparative method using the database of 20 retinal fundus images, which is useful for systems that perform computer-aided diagnosis in clinical practice.

Acknowledgements This research was supported by the National Council of Science and Technology of México under the project: Cátedras-CONACYT 3150-3097.

Appendix

Simulated Annealing code in C programming language used for the Hybrid $UMDA_{Gc}$/SA algorithm proposed in this work.

```
#include "isimulated_annealing.h"
int RealPerturbation(double *Xorig, double *Xpert,int nvar,
 double *xinf,double *xsup, double *params, Irand *seed)
{
  static int i;
  int j=0;
  for (j=0; j<nvar; j++)
      Xpert[j]=Xorig[j];
  if (i<0 || i>nvar-1)
      i=0;
         Xpert[i]=(unifDou(2.0,seed)-1.0)*params[i];
   if (Xpert[i]<xinf[i])
      Xpert[i]=xinf[i];
   else if (Xpert[i]>xsup[i])
      Xpert[i]=xsup[i];
  i++;
return 0;
}


int realSA_UpdateParams(int neps,int generation,
double *ParamsIn, double *ParamsOut, int nvar,
                        realSA_PERTURBATION perType, int Ns,
                        int nprocess,double cu,
                        double *xinf, double *xsup )
{
 double u;
 int i;
 for (i=0; i<nvar; i++)
 {
         u=(ParamsIn[i])/((double)Ns*(double)nprocess);
     if (u>0.6)
             ParamsOut[i]=ParamsOut[i]*(1.0+cu*(u-0.6)/0.4);
     else if(u<0.4)
             ParamsOut[i]=ParamsOut[i]/(1.0+cu*(0.4- u)/0.4);
       if (ParamsOut[i]>(xsup[i]-xinf[i]))
               ParamsOut[i]=(xsup[i]-xinf[i]);
   }
  return 0;
}
```

```
double realSA(FOBJDATA fobjdata,int nvar, double T0,
                      int Neps, int maxeval,double cu,
                      int Ns, double rT, int NT,
                      double eps, EDARESULT *result,int print,
              realSA_PERTURBATION perType,
              double *xinf, double *xsup, Irand *seed)
{
  double **Xlocal, *Xbest, *xeval;
  int nthreads=1;
  double **Xopt,**XBest;
  double *Fopt,*Fbest,fbest,*Flocal,DF,pDF,T=T0,u;
  int **nu;
  int t,s,i,p=1,accept,j,nepsU,neps=0;
  double *ParamsIn,*ParamsOut;
  ///Parameters for perturbing the solution

  ParamsIn=dvector(nvar+1);
  ///Measured parameters for inside use
  ParamsOut=dvector(nvar+1);
  ///Computed parameters for inside use
  xeval=dvector(nvar);
  XBest=dmatrix(nthreads,nvar);
  Xlocal=dmatrix(nthreads,nvar);
  Xopt=dmatrix(nthreads,nvar);
  Flocal=dvector(nthreads);
  Fbest=dvector(nthreads);
  Fopt=dvector(nthreads);
  Xbest=dvector(nvar);
  nu=imatrix(nthreads,nvar);
  ///step size
  for (i=0; i<nvar; i++)
      ParamsOut[i]=(xsup[i]-xinf[i])/2.0;

    for (i=0; i<nvar; i++)
        Xbest[i]=result->xbest[i];
    fbest=result->fbest;
  for (p=0; p<nthreads; p++){
    for (i=0; i<nvar; i++){
        XBest[p][i]=Xbest[i];
        Xopt[p][i]=Xbest[i];
        nu[p][i]=0;
    }
    Fbest[p]=fbest;
    Fopt[p]=fbest;
  }
```

```
///Depending on the perturbation method
int neval=0;
while (neps<Neps && neval<maxeval)
{
for (t=0; t<NT; t++)
{
 for (s=0; s<Ns; s++)
 {
  for (p=0; p<nthreads; p++){
    for (i=0; i<nvar; i++){
      RealPerturbation(XBest[p], Xlocal[p],nvar, xinf,xsup,
                                  ParamsOut,seed);
                Flocal[p]= Fobj(fobjdata,Xlocal[p],nvar);
                (*result).neval++;
                neval++;
                ///Storing the best overall solutions
                if (Flocal[p]>Fopt[p])
                {
                    for (j=0; j<nvar; j++)
                        Xopt[p][j]=Xlocal[p][j];
                    Fopt[p]=Flocal[p];
                }
                ///Metropolis criterion
                DF=Fbest[p]-Flocal[p];
                accept=0;
                if (DF<=0.0) accept=1;
                else{
                    pDF=exp(-DF/T);
                    u=unifDou(1.0,seed);
                    if(u<pDF)
                        accept=1;
                }
                if (accept)
                {
                    for (j=0; j<nvar; j++)
                        XBest[p][j]=Xlocal[p][j];
                    Fbest[p]=Flocal[p];
                    nu[p][i]++;
                }
            }
        }
    }
    }
    ///Best solution found until now
    for (p=0; p<nthreads;p++)
    {
```

```
                    if (Fopt[p]>fbest);
                    {
                        fbest=Fopt[p];
                        for (i=0; i<nvar; i++)
                            Xbest[i]=Xopt[p][i];
                    }
                }
                if (print>=2)
                {
                    printf("t=_%d_fbest=_%lf",t,fbest);
                    for (i=0; i<nvar; i++)
                        printf("%.6g_",Xbest[i]);
                    printf("\n");
                }
                ///Adjusting parameters for the solution perturbation
                for (p=1; p<nthreads; p++)
                    for (i=0; i<nvar; i++)
                        nu[0][i]+=nu[p][i];
                ///Input parametes the number of accepted solutions
                ParamsIn[0]=0.0;
                for (i=0; i<nvar; i++)
                    ParamsIn[0]+=(double)nu[0][i];
                    realSA_UpdateParams(neps,t, ParamsIn, ParamsOut,
                    nvar, perType, Ns,nthreads,cu,xinf, xsup);
                for (p=0; p<nthreads; p++)
                    for (i=0; i<nvar; i++)
                        nu[p][i]=0;
        }
        T=rT*T;
        nepsU=0;
        for (p=0; p<nthreads; p++)
            nepsU+=(int)(fabs(Fbest[p]-fbest)<eps);
        neps+=(int)(nepsU==nthreads);
        for (p=0; p<nthreads; p++){
          for (i=0; i<nvar; i++){
                XBest[p][i]=Xbest[i];
                Xopt[p][i]=Xbest[i];

          }
          Fbest[p]=fbest;
          Fopt[p]=fbest;
          if (print>=1)
            printf("fbest=%.6g_T=%lf_neps=%d\n",fbest,T,neps);
        }
}
```

```
for (i=0; i<nvar;i++)
    result->xbest[i]=Xbest[i];
result->fbest=fbest;

On_FinalizeRutine(1);
free(ParamsIn);
free(ParamsOut);
free_dmatrix(XBest,nthreads) ;
free_dmatrix(Xlocal,nthreads) ;
free_dmatrix(Xopt,nthreads) ;
free(Flocal);
free(Fbest);
free(Fopt);
free(Xbest);
free(xeval);
free_imatrix(nu,nthreads);
return fbest;
}
```

References

1. D. Ballard, Generalizing the Hough transform to detect arbitrary shapes. Pattern Recogn. **13**(2), 111–122 (1981)
2. J. Illingworth, J. Kittler, A survey of the Hough transform. Comput. Vis. Graph. Image Process. **44**(1), 87–116 (1988)
3. V. Leavers, Survey: which Hough transform? Comput. Vis. Graph. Image Process. Image Underst. **58**, 250–264 (1993)
4. R.O. Duda, P.E. Hart, Use of the Hough transformation to detect lines and curves in pictures. Commun. ACM **15**(1), 11–15 (1972)
5. E. Davies, A modified Hough scheme for general circle location. Pattern Recogn. Lett. **7**, 37–43 (1987)
6. D. Ioannou, W. Huda, A. Laine, Circle recognition through a 2D Hough transform and radius histogramming. Image Vis. Comput. **17**, 15–26 (1999)
7. L. Jiang, Efficient randomized Hough transform for circle detection using novel probability sampling and feature points. Optik **123**, 1834–1840 (2012)
8. R. Yip, P. Tam, D. Leung, Modification of Hough transform for circles and ellipses detection using a 2-dimensional array. Pattern Recogn. **25**, 1007–1022 (1992)
9. F. Oloumi, R. Rangayyan, Detection of the temporal arcade in fundus images of the retina using the Hough transform. Conf. Proc. IEEE Eng. Med. Biol. Soc. **1**, 3585–3588 (2009)
10. F. Oloumi, R. Rangayyan, A.L. Ells, Parabolic modeling of the major temporal arcade in retinal fundus images. IEEE Trans. Instrum. Meas. **61**(7), 1825–1838 (2012)
11. N. Chernov, C. Lesort, Least squares fitting of circles. J. Math. Imaging Vis. **23**(3), 239–252 (2005)
12. K. Chung, Y. Huang, A pruning-and-voting strategy to speed up the detection for lines, circles and ellipses. J. Inf. Sci. Eng. **24**(2), 503–520 (2008)
13. L. Xu, E. Oja, P. Kultanen, A new curve detection method: randomized Hough transform (RHT). Pattern Recogn. Lett. **11**(5), 331–338 (1990)
14. X. Zhang, Q. Su, Y. Zhu, Fast algorithm for circle detection using randomized Hough transform. Comput. Eng. Appl. **44**(22), 62–64 (2008)

15. T. Chen, K. Chung, An efficient randomized algorithm for detecting circles. Comput. Vis. Image Underst. **83**(2), 172–191 (2001)
16. L. Jiang, Fast detection of multi-circle with randomized Hough transform. Optim. Lett. **5**(5), 397–400 (2009)
17. H. Cheng, Y. Guo, Y. Zhang, A novel Hough transform based on eliminating particle swarm optimization and its applications. Pattern Recogn. **42**(9), 1959–1969 (2009)
18. E. Cuevas, D. Oliva, D. Zaldivar, M. Perez-Cisneros, H. Sossa, Circle detection using electromagnetism optimization. Inf. Sci. **182**(1), 40–55 (2012)
19. V. Ayala-Ramirez, C.H. Garcia-Capulin, A. Perez-Garcia, R.E. Sanchez-Yanez, Circle detection on images using genetic algorithms. Pattern Recogn. Lett. **27**(6), 652–657 (2006)
20. E. Cuevas, V. Osuna-Enciso, F. Wario, D. Zaldivar, M. Perez-Cisneros, Automatic multiple circle detection based on artificial immune systems. Expert Syst. Appl. **39**, 713–722 (2012)
21. E. Cuevas, D. Zaldivar, M. Perez-Cisneros, M. Ramrez-Ortegon, Circle detection using discrete differential evolution optimization. Pattern Anal. Appl. **14**(1), 93–107 (2011)
22. J. de Jesus Guerrero-Turrubiates, I. Cruz-Aceves, S. Ledesma, J.M. Sierra-Hernandez, J. Velasco, J.G. Avina-Cervantes, M.S. Avila-Garcia, H. Rostro-Gonzalez, R. Rojas-Laguna, Fast parabola detection using estimation of distribution algorithms. Comput. Math. Methods Med. **6494390**, 1–13 (2017)
23. I. Cruz-Aceves, J. Guerrero-Turrubiates, J.M. Sierra-Hernandez, Parametric object detection using estimation of distribution algorithms. Hybrid Intell. Tech. Pattern Anal. Underst. **1**, 69–92 (2017)
24. S. Kirkpatrick, C.D. Gelatt, M.P. Vecchi et al., Optimization by simulated annealing. Science **220**(4598), 671–680 (1983)
25. A. Corana, M. Marchesi, C. Martini, S. Ridella, Minimizing multimodal functions of continuous variables with the simulated annealing algorithm. ACM Trans. Math. Softw. **13**, 262–280 (1987)
26. P. Larrañaga, J. Lozano, *Estimation of Distribution Algorithms: A New Tool for Evolutionary Computation* (Kluwer, Boston, 2002)
27. M. Hauschild, M. Pelikan, An introduction and survey of estimation of distribution algorithms. Swarm Evol. Comput. **1**(3), 111–128 (2011)
28. I. Cruz-Aceves, A. Hernandez-Aguirre, S. Ivvan-Valdez, On the performance of nature inspired algorithms for the automatic segmentation of coronary arteries using Gaussian matched filters. Appl. Soft Comput. **46**, 665–676 (2016)
29. I. Cruz-Aceves, F. Cervantes-Sanchez, A. Hernandez-Aguirre, R. Perez-Rodriguez, A. Ochoa-Zezzatti, A novel Gaussian matched filter based on entropy minimization for automatic segmentation of coronary angiograms. Comput. Electr. Eng. **53**, 263–275 (2016)
30. L. Lozada-Chang, R. Santana, Univariate marginal distribution algorithm dynamics for a class of parametric functions with unitation constraints. Inf. Sci. **181**, 2340–2355 (2011)
31. J.J. Staal, M.D. Abramo, M. Niemeijer, M.A. Viergever, B. van Ginneken, Ridge based vessel segmentation in color images of the retina. IEEE Trans. Med. Imag. **23**(4), 501–509 (2004)
32. M.J. McAuliffe, F.M. Lalonde, D. McGarry, W. Gandler, K. Csaky, B.L. Trus, Medical image processing, analysis & visualization in clinical research, in *Proceedings of the IEEE Symposium on Computer-Based Medical Systems* (2001), pp. 381–388
33. P. Niedfeldt, R. Beard, Recursive RANSAC: multiple signal estimation with outliers. IFAC Proc. Vol. **46**(23), 430–435 (2013). 9th IFAC Symposium on Nonlinear Control Systems

Image Thresholding Based on Fuzzy Particle Swarm Optimization

Anderson Carlos Sousa Santos and Helio Pedrini

Abstract Segmentation is a crucial stage in the image analysis process, whose main purpose is to partition an image into meaningful regions of interest. Thresholding is the simplest image segmentation method, where a global or local threshold value is selected for segmenting pixels into background and foreground regions. However, the determination of a proper threshold value is typically dependent on subjective assumptions or empirical rules. In this work, we propose and analyze an image thresholding technique based on a fuzzy particle swarm optimization. Several images are used in our experiments to show the effectiveness of the developed approach.

Keywords Image thresholding · Particle swarm optimization · Image segmentation · Fuzzy threshold · Fitness function

1 Introduction

Image segmentation is the process of clustering pixels of the image into homogeneous regions based on certain properties, such as brightness, color, and texture. Several applications in the fields of image analysis and computer vision [8, 20, 37] require the segmentation of images, for instance, document analysis, cell counting, brain tumor detection, vehicle license plate recognition, among other problems.

Thresholding [3, 6, 41, 44] is one of the simplest image segmentation approaches, whose main purpose is to extract objects present in the image from its background. Its basic assumption is that the background and objects can be distinguished through their pixel intensity values.

Image thresholding approaches can be classified into bilevel and multilevel techniques. In bilevel thresholding [53, 54], a single threshold value is employed to

A. C. S. Santos (✉) · H. Pedrini
Institute of Computing, University of Campinas, Campinas, SP, Brazil
e-mail: anderson.santos@ic.unicamp.br; helio@ic.unicamp.br

© Springer International Publishing AG, part of Springer Nature 2018
S. Bhattacharyya (ed.), *Hybrid Metaheuristics for Image Analysis*,
https://doi.org/10.1007/978-3-319-77625-5_8

187

partition the image into two portions, one representing the background and another corresponding to the objects. In multilevel thresholding [3, 39], multiple threshold values are used to partition the image into background and objects present in the image.

Entropy maximization [22], intra-class variance minimization [35], histogram shape [35, 49], and Bayesian error minimization [50] are some of the criteria used for the image thresholding approaches available in the literature.

The determination of the best parameters for image segmentation [9, 14, 42] is a serious difficulty found in several thresholding algorithms. In general, the parameters are defined by users based on empirical assumptions. Since the threshold values can significantly influence the position of the objects present in the image, it is crucial to develop efficient mechanisms for selecting the best thresholds.

In this work, the parameters used in different image thresholding techniques are automatically determined by means of a fuzzy particle swarm optimization algorithm. As main contributions, we define an efficient fitness function to evaluate the candidate solutions based on an image quality metric and fuzzy integral, an automatic selection of appropriate parameters for known image thresholding techniques by means of a metaheuristic optimization approach, as well as a combination of particle swarm optimization (PSO) and fuzzy measures, constituting a hybrid metaheuristic method. Experiments are conducted on several images to demonstrate the effectiveness of our method.

The remainder of the chapter is organized as follows. Relevant work associated with image thresholding approaches is briefly reviewed in Sect. 2. The proposed methodology for determining the best parameters for the image thresholding method based on a fuzzy PSO algorithm is described in Sect. 3. Experimental results are reported and analyzed in Sect. 4. Conclusions and directions for future work are presented in Sect. 5.

2   Background

Techniques for image thresholding [6, 15, 20, 51] aim to extract pixels that belong to objects in an image from its background. Thresholding approaches are commonly classified into global and local methods [19, 47], according to their strategies for selecting the threshold values.

Global thresholding methods employ a single threshold value for all the image pixels. This strategy is more adequate when the pixel values corresponding to objects and background are consistent throughout the entire image.

Local thresholding methods use a different threshold adaptively determined for each pixel according to local image properties. This approach is more suitable to adjusting non-uniform lighting conditions or the presence of local shadows in the image.

The following subsections briefly describe some important aspects related to global and local image thresholding techniques.

2.1 Global Thresholding

A straightforward mechanism for global thresholding is to determine an intensity pixel value as a threshold T, where pixel values equal to or above T become 1 (white) and pixel values below T become 0 (black). This strategy can be expressed as

$$\mathbf{g}(x, y) = \begin{cases} 1, & \text{if } \mathbf{f}(x, y) \geq T \\ 0, & \text{otherwise} \end{cases} \tag{1}$$

where (x, y) corresponds to a pixel position, whereas $\mathbf{f}$ and $\mathbf{g}$ are the original image and the transformed image, respectively.

Otsu [35] developed an automatic method for globally grouping pixels of the image into background and foreground regions. The algorithm exhaustively searches for the threshold that maximizes the inter-class variance (the variance between the classes) or, conversely, minimizes the intra-class variance (that is, the variance within the class).

The inter-class variance can be defined as

$$\sigma^2_{\text{between}}(T) = n_B(T)\, n_F(T)[\mu_B(T) - \mu_F(T)]^2, \tag{2}$$

where $\mu_B(T) = \sum_{i=0}^{T-1} i\, p(i)/n_B(T)$ and $\mu_F(T) = \sum_{i=T}^{L-1} i\, p(i)/n_F(T)$. The intensity levels of the image pixels are in the range $[0, L - 1]$.

The intra-class variance can be defined as the weighted sum of the variances of each class

$$\sigma^2_{\text{within}}(T) = n_B(T)\, \sigma^2_B(T) + n_F(T)\, \sigma^2_F(T), \tag{3}$$

where $n_B(T) = \sum_{i=0}^{T-1} p(i)$ and $n_F(T) = \sum_{i=T}^{L-1} p(i)$. The value $\sigma^2_B(T)$ corresponds to the variance of the pixels in the background (below the threshold), whereas $\sigma^2_F(T)$ corresponds to the variance of the pixels in the foreground (above the threshold).

2.2 Local Thresholding

Bernsen [7] developed a method that computes the local threshold value based on the mean value of the minimum and maximum pixel intensities within a neighborhood, expressed as

$$T(x, y) = (z_{\min} + z_{\max})/2 \tag{4}$$

where $z_{\min}$ and $z_{\max}$ are the minimum and maximum intensity values, respectively, within an $n \times n$ region centered at (x, y).

Niblack [33] proposed a method that calculates the threshold value for each pixel (x, y) based on the local mean and standard deviation of pixel intensities within a neighborhood, expressed as

$$T(x, y) = \mu(x, y) + k\,\sigma(x, y) \tag{5}$$

where $\mu(x, y)$ and $\sigma(x, y)$ are the local mean and standard deviation, respectively, within an $n \times n$ region of pixel (x, y). The value k is used to adjust the fraction of total pixels present in the foreground object.

Sauvola and Pietaksinen [40] developed a method that calculates the threshold value for each pixel (x, y) as

$$T(x, y) = \mu(x, y)\left[1 + k\left(\frac{\sigma(x, y)}{R} - 1\right)\right] \tag{6}$$

where $\mu(x, y)$ and $\sigma(x, y)$ are local mean and standard deviation, respectively, within an $n \times n$ region of pixel (x, y). Sauvola and Pietaksinen suggested the values $k = 0.5$ and $R = 128$.

2.3 Metaheuristics for Image Thresholding

There are some metaheuristic approaches available in the literature for image thresholding purposes. Oliva et al. [34] utilized a harmony search evolutionary method for multilevel thresholding. Manikandan et al. [31] used real coded genetic algorithms for brain image segmentation with thresholding. Malisia and Tizhoosh [30] proposed the use of Ant Colony Optimization to generate another image representation, called a pheromone matrix, which is used along with the original normalized image for the classification of black-and-white pixels using K-means clustering. Liang et al. [27] also employed Ant Colony Optimization for multilevel thresholding applied to Otsu's criteria [35].

A differential evolution algorithm was applied for thresholding by Sarkar and Das [38]. Charansiriphasian et al. [12] proposed an improved version of the approach developed by Sarkar and Das [38].

Brajevic and Tuba [10] defined a threshold for each pixel using Cuckoo Search, Firefly Algorithm, and Kapur's criteria [22] as the objective function. Alihodzic and Tuba [4] proposed a modification of the Bat Algorithm to perform a search for thresholds using both Kapur's and Otsu's criteria.

PSO has been adopted in some image thresholding approaches. Yin [52] applied PSO for further searching, following a minimization process using recursive programming. A modified version of PSO with run-time adaptive population size was proposed by Liu et al. [28] for multilevel image thresholding purposes. Liu et al. [29] improved the PSO algorithm for thresholding with adaptive inertia weight.

Kurban et al. [25] compared the results of PSO against other nature inspired and evolutionary methods for image thresholding. Li et al. [26] applied a quantum-behaved PSO for medical image segmentation with Otsu's thresholding criteria.

Differently from the methods available in the literature, our proposal aims to optimize the parameters of local adaptive thresholding methods instead of the threshold value itself. Most of the metaheuristics in the literature are used to estimate the actual threshold value. In order to avoid a prohibitive search, they work on a global level. Another distinction is the objective function used in our method. Most approaches use standard theoretical criteria which do not correspond to true quality results, but rather histogram partition. The method developed and analyzed in this chapter introduces a new criterion based on qualitative measures.

3 Methodology

In our methodology, an algorithm for PSO is applied to obtain the best arguments for two of the most common local thresholding algorithms: Niblack's [33] and Sauvola's method [40]. We propose a fitness function for the PSO algorithm based on image quality similarity using a fuzzy formulation of the Structural Similarity Index Measure (SSIM) [48].

Figure 1 illustrates the most relevant stages of the proposed approach. The following subsections describe each step in more detail.

3.1 Particle Swarm Optimization

Swarm intelligence [2, 36] refers to the decentralized and collective behavior that allows for a best solution in solving a problem. Artificial swarm intelligence approaches were created as analogies to living beings that perform a joint effort as a group to achieve a certain goal. In nature, an example is a colony of ants or bees [17].

PSO [23, 43] was inspired by the organized movement in a flock of birds or school of fish. It stands out as one of these swarm intelligence techniques. The problem is solved by improving a candidate solution according to a specified quality measure, where a population of particles (the swarm) moves around the problem search space of real values, such that the movement of particles depends on their local best known positions and their best positions in the search space [13].

As a metaheuristic, PSO does not guarantee an optimal solution. Few assumptions are made about the problem under investigation to search for large spaces by starting with a sample of random solutions. Nevertheless, it is suitable for local image thresholding optimization since the problem has a large search space and is not differentiable.

PSO was chosen in this work due to the various advantages associated with this approach. Its convergence is fast for our setup with only one or two variables. The

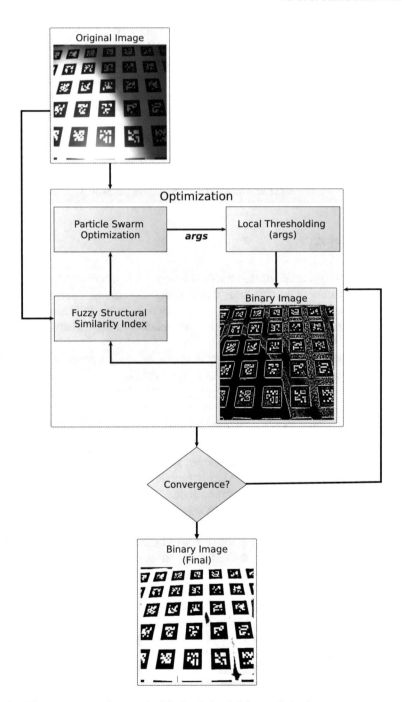

Fig. 1 Main components of our method for local thresholding optimization

technique is straightforward to implement and easily parallelized. Moreover, PSO is simple to set up when only a few parameters are required to be estimated [1].

Although there are many variations of PSO available in the literature, the fundamental concept associated with this optimization consists in an initial random population (called a swarm) of candidate solutions (called particles), each one having the parameters to be optimized [11]. At each iteration, every particle adjusts its velocity vector and the best known position of the swarm is updated according to a certain fitness function.

The PSO algorithm stores and progressively replaces the best parameters of each particle [46], as well as the particle that best fits the parameters. The process continues until a predefined convergence criterion is reached. Algorithm 1 presents the PSO version used throughout this work.

The parameter N refers to the number of particles and is defined based on the problem's difficulty and the solution's dimensionality.

The other parameters are associated with the velocity update equation shown in line 7 of Algorithm 1, where ω represents an inertia weight used to control the impact of the previous velocity value. The coefficients ϕ_p and ϕ_g serve as cognitive and social influence, controlling the effect of personal and the swarm's best position on the direction of the movement, respectively.[1]

Algorithm 1: Particle swarm optimization

---

input : swarm size N, number of variables D, inertia ω, acceleration coefficients ϕ_p and ϕ_g
output: best solution

---

1 Create an array x for particle positions with D dimensions
2 Initialize particle velocities: $v \leftarrow 0$
3 **while** *convergence is not reached* **do**
4   **for** $i \in 0, \dots, N$ **do**
5   Generate random vectors $U_i(\phi_p)$ and $U_i(\phi_g)$
6   Compute particle velocities
7              $v_i = \omega * v_i + U_i(\phi_p) * (p_i - x_i) + U_i(\phi_g) * (g - x_i)$
8   Update particle positions
9              $x_i = x_i + v_i$
10   Compute objective function f
11   **if** $f(x_i) < f(p_i)$ **then**
12   Update particle's best position
13                 $p_i \leftarrow x_i$
14   **if** $f(p_i) < f(g)$ **then**
15   Update swarm's best position
16                     $g \leftarrow p_i$
17   **end**
18   **end**
19 **end**
20 **end**
21 **return** *best particle position* g.

---

[1] $U_i(\phi)$ is an array of random numbers uniformly distributed in the range $[0, \phi]$.

Our implementation has multiple convergence criteria that stops the PSO algorithm if:

- number of iterations > 200
- best objective value stalls for 30 iterations
- change in best objective value $< 10^{-8}$ or
- change in best position $< 10^{-8}$.

Since the parameters of our local thresholding method correspond to the optimization goal, here we consider the entire process of computing a binary image as the objective function (line 10, Algorithm 1) and compare it to the original image through a quality measure. Section 3.2 presents the details of the quality metric.

A common parameter in all local thresholding methods is the neighborhood in which they operate. Since it consists of an integer and has a limited search space, we search for the best region size iteratively by computing a PSO for each, as shown in Algorithm 2.

The search is performed from a vicinity of 1 until 25 with a step of 2, whereas the local threshold arguments and neighborhood size corresponding to the minimum value of the objective function are returned.

3.2   Fuzzy Structural Similarity Index

As mentioned in Sect. 3.1, the objective function comprehends the application of a local threshold method and the comparison between the resulting binary image and original gray scale image. To obtain a good quality measure for the binarization process, we propose the use of a fuzzy version of the SSIM.

Algorithm 2: Neighborhood search
___

input : original image img
output: best solution

1 $minVal \leftarrow +\infty$
2 **for** $n \in 1, 3, 5, \ldots, 23, 25$ **do**
3 $\quad args \leftarrow PSO$(Algorithm 1)
4 $\quad binaryImg \leftarrow localThreshold(img, args, n)$
5 $\quad fval \leftarrow fuzzyDSSIM(img, binaryImg)$
6 $\quad$ **if** $fval < minVal$ **then**
7 $\quad\quad minVal \leftarrow fval$
8 $\quad\quad minArgs \leftarrow args$
9 $\quad\quad minN \leftarrow n$
10 $\quad$ **end**
11 **end**
12 **return** $minArgs$ and $minN$.

The SSIM [48] is a quality measure calculated originally between two gray scale images $\mathbf{f}$ and $\mathbf{g}$. This is calculated for various image windows $\mathbf{x}$ and $\mathbf{y}$, expressed as

$$\mathrm{SSIM}(\mathbf{x}, \mathbf{y}) = \frac{(2\mu_x \mu_y + C_1)(2\sigma_{xy} + C_2)}{(\mu_x^2 + \mu_y^2 + C_1)(\sigma_x^2 + \sigma_y^2 + C_2)} \tag{7}$$

where μ_x is the mean of $\mathbf{x}$, μ_y is the mean of $\mathbf{y}$, σ_x^2 is the variance of $\mathbf{x}$, σ_y^2 is the variance of $\mathbf{y}$, and σ_{xy} is the covariance of $\mathbf{x}$ and $\mathbf{y}$, C_1 and C_2 are constants that stabilize the equation ($C_1 = 0.01 \times 255^2$ and $C_2 = 0.03 \times 255^2$).

The final SSIM measure is the average of all windows in each pixel and varies in the range of $[-1, 1]$, such that the closer it is to 1, the more similar the images $\mathbf{f}$ and $\mathbf{g}$ are.

The formula presented in Eq. (7) can be decomposed into three individual functions:

$$l(x, y) = \frac{2\mu_x \mu_y + C_1}{\mu_x^2 + \mu_y^2 + C_1} \tag{8}$$

$$c(x, y) = \frac{2\sigma_x \sigma_y + C_2}{\sigma_x^2 + \sigma_y^2 + C_2} \tag{9}$$

$$s(x, y) = \frac{\sigma_x y + C_3}{\sigma_x \sigma_y + C_3} \tag{10}$$

related to luminance (8), contrast (9), and structure (10).

Our main modification relies on the computation of the overall index measure, replacing the average metric by the Sugeno Fuzzy Integral (SFI) [21, 45]. This is based on the fuzzy measure theory [24, 32] that considers the subjectiveness of information sources through measures that quantify the *a priori* importance of an individual source and its possible coalitions.

More formally, coefficients $\mu(A_j)$ are defined for all subsets of the set of integrands (χ) in the range $[0, 1]$ and must meet the monotonicity condition given in Eq. (11).

$$A_j \subset A_k \implies \mu(A_j) \leq \mu(A_k) \quad \forall A_j, A_k \in \chi \tag{11}$$

A fuzzy integral [5, 16] aggregates the information according to the fuzzy measure. The computation of the SFI [21, 45] uses minimum ($\wedge$) and maximum ($\vee$) operators to quantify the level of agreement between the different sources, as shown in Eq. (12).

$$S_\mu[h_1(x_i), \ldots, h_n(x_n)] = \bigvee_{i=1}^{n} [h_{(i)}(x_i) \wedge \mu(A_{(i)})] \tag{12}$$

Algorithm 3: Fuzzy DSSIM

---

    input : images A and B
    output: dissimilarity measure

1 $X \leftarrow SSIM(A, B)$
2 $X \leftarrow abs(X)$
3 $X^s \leftarrow reverseSort(X)$
4 $min \leftarrow \emptyset$
5 **for** $i \in 1, \ldots, |X^s|$ **do**
6      $\mu_i \leftarrow \frac{i}{|X^s|}$
7      $min \leftarrow minimum(X_i^s, \mu_i)$
8 **end**
9 $fuzzySSIM \leftarrow max(min)$
10 **return** $1 - fuzzySSIM$

---

where the sub-index $_{(i)}$ corresponds to a previous decreasing sort on the integrands, where $h_{(1)}(x_1)$ is the source with the highest value, and $A_{(k)}$ is the subset with the k highest values, such that $A_{(n)} = \chi$.

In our problem, the sources of information are the structural similarities between regions represented as an image X, where each pixel maintains the SSIM value of the region around it. Nonetheless, in order to use the fuzzy integral, this similarity must represent a fuzzy membership. To accomplish that, we used the absolute value of SSIM since the negative value comes from Eq. (10), which means a difference only in luminance and contrast, but a correlation in structure.

We set the fuzzy measures as probability measures, similarly to Gao et al. [18]. Thus, the fuzzy measure is defined as given in Eq. (13).

$$\mu(A_{(i)}) = \frac{|A_{(i)}|}{N} \tag{13}$$

where $|A_{(i)}|$ is the cardinality of the set.

Algorithm 3 resumes all the steps for the computation of the quality measure between images. Since PSO performs a minimization optimization, it is important to observe that the output of the algorithm is a dissimilarity metric (DSSIM), defined as one minus the similarity measure.

4 Results

To validate and analyze our methodology, experiments were conducted on several gray scale images. Due to space limitation, results for four images (shown in Fig. 2) are presented in this section.

We compared Niblack's [33] and Sauvola's methods [40] under our framework and with default parameters, compared their results to Otsu's global thresholding approach [35] and assessed the impact of the Fuzzy Structural Dissimilarity Index as the objective function.

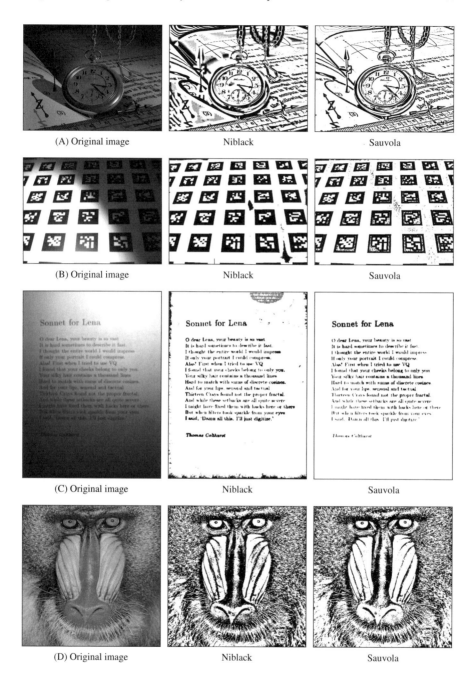

Fig. 2 Results for input images after estimating the best parameters for the thresholding methods and Otsu's global thresholding approach

4.1 Parameter Setup

For the PSO, the inertia (ω) and the acceleration coefficients ϕ_p and ϕ_g were all set to 0.5. The maximum number of iterations was set to 200 and the maximum stall iterations was set to 30. The size of the swarm was defined as 20 for optimization of Niblack's method and 100 for Sauvola's method, since the latter has a large search space.

Furthermore, the search space size is limited by the lower and upper bounds. For the k parameter shared in both thresholding methods (Eqs. (5) and (6)), its range varied from -10 to 10. In Sauvola's method, the R parameter was defined between 1 and 255.

The default parameters were used in the SSIM [48]. This was computed in each pixel with a window of 7×7 and a uniform filter.

4.2 Results

Figure 2 shows the results of the application of two local thresholding procedures using the estimated parameters found with our optimization method. The exact arguments used are reported in Table 1, as well as the value of Fuzzy SSIM between the binary and original image.

Table 1 Best estimated arguments and objective value

Image	Method	Arguments	Fuzzy DSSIM
A	Niblack	$n = 19$	0.786697
		$k = -0.098298$	
	Sauvola	$n = 9$	0.78001
		$k = 0.022729$	
		$R = 74.967747$	
B	Niblack	$n = 23$	0.684108
		$k = -0.002992$	
	Sauvola	$n = 7$	0.681683
		$k = 0.020232$	
		$R = 115.061960$	
C	Niblack	$n = 21$	0.83455
		$k = -0.936996$	
	Sauvola	$n = 13$	0.833354
		$k = 0.287540$	
		$R = 25.102691$	
D	Niblack	$n = 23$	0.645525
		$k = -0.101720$	
	Sauvola	$n = 23$	0.644344
		$k = 0.065963$	
		$R = 193.819038$	

It is noticeable from Fig. 2 that Sauvola's method presented noise-free images, especially for image C. On the other hand, it is difficult to determine which method presented a better quality image for image B. The results for Niblack's method show some large unwanted blobs, whereas there is more degradation and scatter noise for Sauvola's method. The choice will depend on the preferred post-processing technique.

The values of Fuzzy DSSIM, reported in Table 1, are considered high and closer to one than zero, indicating that the images are very different. However, this was due to the different nature of the images (gray scale versus binary). The relative quality perception is preserved since Sauvola's method presents a smaller value of Fuzzy DSSIM for all images, although the difference is tiny in absolute values.

Another interesting aspect in Table 1 is the precision of the arguments found by the method, which would be very unlikely for a human to infer. The values obtained differ significantly among images and the local thresholding method used, with the exception of the neighborhood size in some cases. This shows the difficulty in working with such methods without some optimization, as well as the adaptive power of our methodology.

To demonstrate the advantages of a local thresholding method with the parameters set properly over a blunt guess or a global thresholding, Figs. 3 and 4 show the results for Otsu's global thresholding [35], the local methods with default configuration, and through our optimization strategy. The default settings for Niblack's method consist in a neighborhood of 5 and $k = -0.2$, whereas a neighborhood of 5, $k = 0.5$, and $R = 128$ applies to Sauvola's method.

It is possible to observe that the resulting images with an adaptive threshold can be worse than a global threshold given an erroneous configuration. Sauvola's method presents inferior results with its default values despite being considered better in our framework. This is understandable, since that method requires one extra parameter when compared with Niblack's, which increases the combinatorial possibilities, making it hard to guess the correct combination.

To assess the impact of the fuzzy version of the Structural Dissimilarity Index as the fitness function, we evaluate the PSO using the regular DSSIM as the objective function. Figures 5 and 6 show the results for both regular and Fuzzy DSSIM for Niblack's and Sauvola's method, respectively.

For some cases, such as image B, there is no perceptible difference among the resulting binary images, except for image C, where a huge difference can be noticed for both methods. The result using a fuzzy integral presented more consistent and noise-free images as expected, since it reinforces the notion of overall agreement compared with the average, which is susceptible to outliers. This can be noticed in Table 2, where the arguments and values for DSSIM are reported. By analyzing the values with Fuzzy DSSIM in Table 1, it is possible to observe that DSSIM outputs lower values, indicating that it is more permissive and flexible in terms of similarity.

From the arguments found through the optimized thresholding methods, except for the neighborhood in image B, there was no coincidence between the two optimization strategies, although some images presented a similar visual result. This

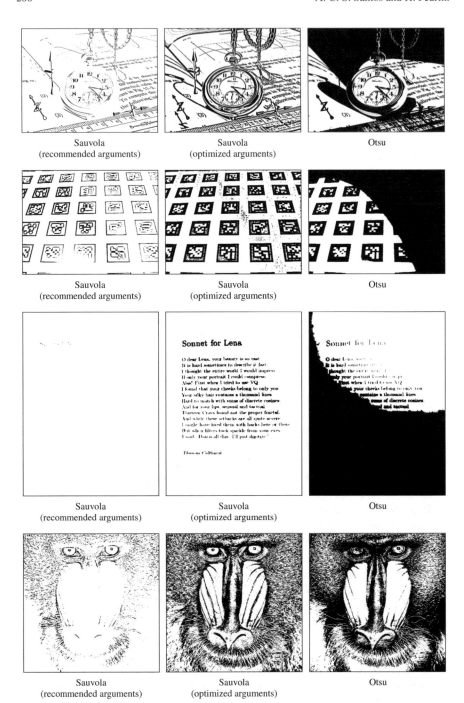

Fig. 3 Comparison against optimized Niblack's thresholding method, its standard version, and Otsu's global thresholding approach

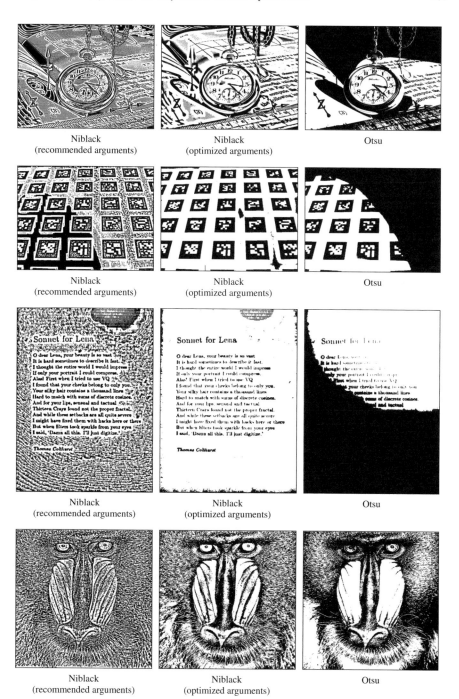

Fig. 4 Comparison against optimized Sauvola's thresholding method, its standard version, and Otsu's global thresholding approach

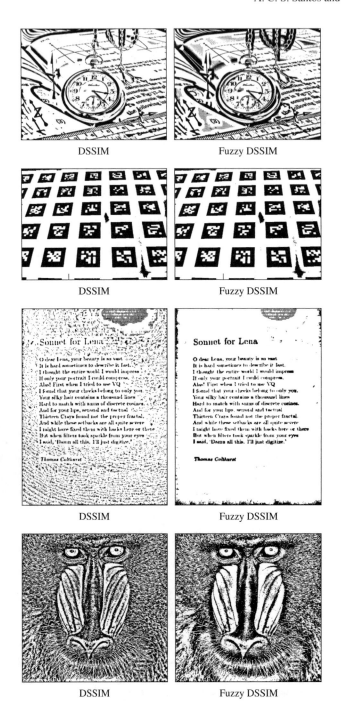

DSSIM   Fuzzy DSSIM

DSSIM   Fuzzy DSSIM

DSSIM   Fuzzy DSSIM

DSSIM   Fuzzy DSSIM

Fig. 5 Comparison between the use of standard DSSIM and Fuzzy DSSIM as an objective function for Niblack's thresholding method

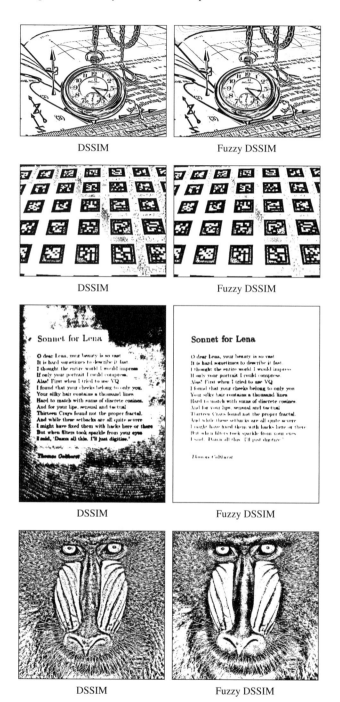

Fig. 6 Comparison between the use of standard DSSIM and Fuzzy DSSIM as an objective function for Sauvola's thresholding method

Table 2 Best estimated arguments and objective value using standard DSSIM

Image	Method	Arguments	DSSIM
A	Niblack	$n = 23$	0.442512
		$k = -0.330241$	
	Sauvola	$n = 7$	0.434645
		$k = 0.037380$	
		$R = 15.690637$	
B	Niblack	$n = 23$	0.352743
		$k = -0.065023$	
	Sauvola	$n = 7$	0.350361
		$k = 0.011847$	
		$R = 170.203209$	
C	Niblack	$n = 5$	0.469049
		$k = -0.740884$	
	Sauvola	$n = 19$	0.47028
		$k = -0.042139$	
		$R = 5.226428$	
D	Niblack	$n = 9$	0.364504
		$k = -0.086095$	
	Sauvola	$n = 9$	0.36346
		$k = 0.033138$	
		$R = 238.415270$	

reinforces how fine tuned the local thresholding methods need to be in order to generate more adequate results.

5 Conclusions

The selection of threshold values for image segmentation is typically performed manually based on empirical assumptions; however, this process may significantly affect the quality of the final result.

This chapter has presented and discussed the application of a PSO, with a fuzzy formulation of the image quality metric as an objective function, to two image thresholding techniques. Experiments were conducted on different images. The results were analyzed and compared to traditional uses of thresholding approaches.

The automatic threshold values estimated by the algorithms, according to a fitness function based on fuzzy image quality, were able to generate competitive results that in many cases shown a better visual result. The proposed method can be used as a preprocessing step in other tasks, such as object location and image interpretation.

As directions for future work, we intend to analyze the improvement of different image processing techniques, for instance, filtering, registration, and compression. Furthermore, other optimization approaches could benefit from our proposed fuzzy fitness function for image quality to improve the selection of image segmentation parameters.

Acknowledgements The authors are thankful to São Paulo Research Foundation (grant FAPESP #2014/12236-1) and the Brazilian Council for Scientific and Technological Development (grant CNPq #305169/2015-7 and scholarship #141647/2017-5) for their financial support.

References

1. M.N. Ab Wahab, S. Nefti-Meziani, A. Atyabi, A comprehensive review of swarm optimization algorithms. PloS One **10**(5), e0122827 (2015)
2. A. Abraham, H. Guo, H. Liu, Swarm intelligence: foundations, perspectives and applications, in *Swarm Intelligent Systems* (Springer, Berlin, 2006), pp. 3–25
3. M. Ali, C.W. Ahn, M. Pant, Multi-level image thresholding by synergetic differential evolution. Appl. Soft Comput. **17**, 1–11 (2014)
4. A. Alihodzic, M. Tuba, Improved bat algorithm applied to multilevel image thresholding. Sci. World J. **2014**, 16 pp. (2014)
5. D.T. Anderson, T.C. Havens, C. Wagner, J.M. Keller, M.F. Anderson, D.J. Wescott, Sugeno fuzzy integral generalizations for sub-normal fuzzy set-valued inputs, in *IEEE International Conference on Fuzzy Systems* (2012), pp. 1–8
6. M. Beauchemin, Image thresholding based on semivariance. Pattern Recogn. Lett. **34**(5), 456–462 (2013)
7. J. Bernsen, Dynamic thresholding of grey-level images, in *6th International Conference on Pattern Recognition*, Berlin (1986), pp. 1251–1255
8. B. Bhanu, S. Lee, *Genetic Learning for Adaptive Image Segmentation*, vol. 287 (Springer Science & Business Media, Berlin, 2012)
9. B. Bhanu, S. Lee, S. Das, Adaptive image segmentation using genetic and hybrid search methods. IEEE Trans. Aerosp. Electron. Syst. **31**(4), 1268–1291 (1995)
10. I. Brajevic, M. Tuba, Cuckoo search and firefly algorithm applied to multilevel image thresholding, in *Cuckoo Search and Firefly Algorithm* (Springer, Berlin, 2014), pp. 115–139
11. D. Bratton, J. Kennedy, Defining a standard for particle swarm optimization, in *IEEE Swarm Intelligence Symposium* (IEEE, New York, 2007), pp. 120–127
12. K. Charansiriphaisan, S. Chiewchanwattana, K. Sunat, A global multilevel thresholding using differential evolution approach. Math. Probl. Eng. **2014**, 23 pp. (2014)
13. M. Clerc, *Particle Swarm Optimization* (Wiley, New York, 2010)
14. J. D'Avy, W.-W. Hsu, C.-H. Chen, A. Koschan, M. Abidi, An efficient method for optimizing segmentation parameters, in *Emerging Technologies in Intelligent Applications for Image and Video Processing* (IGI Global, Hershey, 2016), pp. 29–47
15. A. Dirami, K. Hammouche, M. Diaf, P. Siarry, Fast multilevel thresholding for image segmentation through a multiphase level set method. Signal Process. **93**(1), 139–153 (2013)
16. D.J. Dubois, *Fuzzy Sets and Systems: Theory and Applications*, vol. 144 (Academic, New York, 1980)
17. R.C. Eberhart, Y. Shi, J. Kennedy, *Swarm Intelligence* (Elsevier, Amsterdam, 2001)
18. X. Gao, T. Wang, J. Li, A content-based image quality metric, in *Rough Sets, Fuzzy Sets, Data Mining, and Granular Computing* (Springer, Berlin, 2005), pp. 231–240

19. C.A. Glasbey, An analysis of histogram-based thresholding algorithms. Graph. Models Image Process. **55**(6), 532–537 (1993)
20. R. Gonzalez, R. Woods, *Digital Image Processing Using Matlab* (McGraw Hill Education, New York, 2010)
21. M. Grabisch, M. Sugeno, T. Murofushi, *Fuzzy Measures and Integrals: Theory and Applications* (Springer, New York, 2000)
22. J.N. Kapur, P.K. Sahoo, A.K. Wong, A new method for gray-level picture thresholding using the entropy of the histogram. Comput. Vis. Graph. Image Process. **29**(3), 273–285 (1985)
23. J. Kennedy, R. Eberhart, Particle swarm optimization, in *IEEE International Conference on Neural Networks*, Perth, 1995, pp. 1942–1948
24. G. Klir, B. Yuan, *Fuzzy Sets and Fuzzy Logic*, vol. 4 (Prentice Hall, Princeton, 1995)
25. T. Kurban, P. Civicioglu, R. Kurban, E. Besdok, Comparison of evolutionary and swarm based computational techniques for multilevel color image thresholding. Appl. Soft Comput. **23**, 128–143 (2014)
26. Y. Li, L. Jiao, R. Shang, R. Stolkin, Dynamic-context cooperative quantum-behaved particle swarm optimization based on multilevel thresholding applied to medical image segmentation. Inform. Sci. **294**, 408–422 (2015)
27. Y.-C. Liang, A.H.-L. Chen, C.-C. Chyu, Application of a hybrid ant colony optimization for the multilevel thresholding in image processing, in *International Conference on Neural Information Processing* (Springer, Berlin, 2006), pp. 1183–1192
28. Y. Liu, C. Mu, W. Kou, Optimal multilevel thresholding using the modified adaptive particle swarm optimization. Int. J. Digital Content Technol. Appl. **6**(15), 208–219 (2012)
29. Y. Liu, C. Mu, W. Kou, J. Liu, Modified particle swarm optimization-based multilevel thresholding for image segmentation. Soft Comput. **19**(5), 1311–1327 (2015)
30. A.R. Malisia, H.R. Tizhoosh, Image thresholding using ant colony optimization, in *3rd Canadian Conference on Computer and Robot Vision* (2006)
31. S. Manikandan, K. Ramar, M.W. Iruthayarajan, K. Srinivasagan, Multilevel thresholding for segmentation of medical brain images using real coded genetic algorithm. Measurement **47**, 558–568 (2014)
32. T. Murofushi, M. Sugeno, Fuzzy measures and fuzzy integrals, in *Fuzzy Measures and Integrals – Theory and Applications*, ed. by M. Grabisch, T. Murofushi, M. Sugeno (Physica, Heidelberg, 2000), pp. 3–41
33. W. Niblack, *An Introduction to Digital Image Processing* (Prentice Hall, Princeton, 1986)
34. D. Oliva, E. Cuevas, G. Pajares, D. Zaldivar, M. Perez-Cisneros, Multilevel thresholding segmentation based on harmony search optimization. J. Appl. Math. **2013** (2013)
35. N. Otsu, A threshold selection method from gray-level histograms. IEEE Trans. Syst. Man Cybern. **9**(1), 62–66 (1979)
36. R. Poli, J. Kennedy, T. Blackwell, Particle swarm optimization. Swarm Intell. **1**(1), 33–57 (2007)
37. A. Rosenfeld, *Multiresolution Image Processing and Analysis*, vol. 12 (Springer Science & Business Media, Berlin, 2013)
38. S. Sarkar, S. Das, Multilevel image thresholding based on 2D histogram and maximum tsallis entropy: a differential evolution approach. IEEE Trans. Image Process. **22**(12), 4788–4797 (2013)
39. S. Sarkar, S. Das, S.S. Chaudhuri, A multilevel color image thresholding scheme based on minimum cross entropy and differential evolution. Pattern Recogn. Lett. **54**, 27–35 (2015)
40. J. Sauvola, M. Pietaksinen, Adaptive document image binarization. Pattern Recogn. **33**, 225–236 (2000)
41. M. Sezgin, Survey over image thresholding techniques and quantitative performance evaluation. J. Electron. Imaging **13**(1), 146–168 (2004)
42. S. Shen, W. Sandham, M. Granat, A. Sterr, MRI fuzzy segmentation of brain tissue using neighborhood attraction with neural-network optimization. IEEE Trans. Inform. Technol. Biomed. **9**(3), 459–467 (2005)

43. Y. Shi, R. Eberhart, A modified particle swarm optimizer, in *IEEE International Conference on Evolutionary Computation* (IEEE, New York, 1998), pp. 69–73
44. A. Singla, S. Patra, A context sensitive thresholding technique for automatic image segmentation, in *Computational Intelligence in Data Mining*, ed. by L.C. Jain, H.S. Behera, J.K. Mandal, D.P. Mohapatra. Smart Innovation, Systems and Technologies, vol. 32, (Springer India, New Delhi, 2015), pp. 19–25
45. M. Sugeno, Theory of fuzzy integrals and its applications. Ph.D. thesis, Tokyo Institute of Technology, 1974
46. I.C. Trelea, The particle swarm optimization algorithm: convergence analysis and parameter selection. Inform. Process. Lett. **85**(6), 317–325 (2003)
47. B.D. Trier, A.K. Jain, Goal-directed evaluation of binarization methods. IEEE Trans. Pattern Anal. Mach. Intell. **17**(12), 1191–1201 (1995)
48. Z. Wang, A.C. Bovik, H.R. Sheikh, E.P. Simoncelli, Image quality assessment: from error visibility to structural similarity. IEEE Trans. Image Process. **13**(4), 600–612 (2004)
49. J.S. Weszka, R.N. Nagel, A. Rosenfeld, A threshold selection technique. IEEE Trans. Comput. **C-23**, 1322–1326 (1974)
50. Q.-Z. Ye, P.-E. Danielsson, On minimum error thresholding and its implementations. Pattern Recogn. Lett. **7**(4), 201–206 (1988)
51. Z. Ye, Z. Hu, X. Lai, H. Chen, Image segmentation using thresholding and swarm intelligence. J. Softw. **7**(5), 1074–1082 (2012)
52. P.-Y. Yin, Multilevel minimum cross entropy threshold selection based on particle swarm optimization. Appl. Math. Comput. **184**(2), 503–513 (2007)
53. Y. Zou, H. Liu, E. Song, Z. Huang, Image bilevel thresholding based on multiscale gradient multiplication. Comput. Electr. Eng. **38**(4), 853–861 (2012)
54. Y. Zou, H. Liu, Q. Zhang, Image bilevel thresholding based on stable transition region set. Digital Signal Process. **23**(1), 126–141 (2013)

Hybrid Metaheuristics Applied to Image Reconstruction for an Electrical Impedance Tomography Prototype

Wellington Pinheiro dos Santos, Ricardo Emmanuel de Souza, Valter Augusto de Freitas Barbosa, Reiga Ramalho Ribeiro, Allan Rivalles Souza Feitosa, Victor Luiz Bezerra Araújo da Silva, David Edson Ribeiro, Rafaela Covello Freitas, Juliana Carneiro Gomes, Natália Souza Soares, Manoela Paschoal de Medeiros Lima, Rodrigo Beltrão Valença, Rodrigo Luiz Tomio Ogava, and Ítalo José do Nascimento Silva Araújo Dias

Abstract Evolutionary computation has much scope for solving several important practical applications. However, sometimes they return only marginal performance, related to inappropriate selection of various parameters (tuning), inadequate representation, the number of iterations and stop criteria, and so on. For these cases, hybridization could be a reasonable way to improve the performance of algorithms. Electrical impedance tomography (EIT) is a non-invasive imaging technique free of ionizing radiation. EIT image reconstruction is considered an ill-posed problem and, therefore, its results are dependent on dynamics and constraints of reconstruction algorithms. The use of evolutionary and bioinspired techniques to reconstruct EIT images has been taking place in the reconstruction algorithm area with promising qualitative results. In this chapter, we discuss the implementation of evolutionary and bioinspired algorithms and its hybridizations to EIT image reconstruction. Quantitative and qualitative analyses of the results demonstrate that hybrid algorithms, here considered, in general, obtain more coherent anatomical images than canonical and non-hybrid algorithms.

Keywords Metaheuristics · Hybridization · Particle swarm optimization · Differential evolution · Fish school search · Density based on fish school search · Electrical impedance tomography · Image reconstruction

W. P. dos Santos (✉) · R. E. de Souza · V. A. de Freitas Barbosa · R. R. Ribeiro · A. R. S. Feitosa · D. E. Ribeiro · J. C. Gomes · N. S. Soares · M. P. de Medeiros Lima · R. B. Valença · R. L. T. Ogava · Í. J. d. N. S. A. Dias
Universidade Federal de Pernambuco, Recife, Brazil
e-mail: wellington.santos@ufpe.br

V. L. B. A. da Silva · R. C. Freitas
Escola Politécnica da Universidade de Pernambuco, Recife, Pernambuco, Brazil

© Springer International Publishing AG, part of Springer Nature 2018
S. Bhattacharyya (ed.), *Hybrid Metaheuristics for Image Analysis*,
https://doi.org/10.1007/978-3-319-77625-5_9

1 Introduction

From the several areas of computational intelligence, evolutionary computation has been emerging as one of the most important sets of problems for solving methodologies and tools in many fields of engineering and computing [1–5]. When compared to other optimization techniques, learning processes based on population, self-adaptation, and robustness appear as fundamental aspects of evolutionary algorithms in comparison to other optimization techniques [1–3].

Despite the large acceptance of evolutionary computation to solve several important applications in several fields such as engineering, e-commerce, business, economy, and health, they tend to return marginal performance [1–3]. Such limitation is related to the large numbers of parameters being selected (the tuning problem), inadequate data representation, the number of iterations, and stop criteria. Nevertheless, according to the No Free Lunch theorem, it is not possible to find the best optimization algorithm for solving all problems with uniform performance (i.e. for all algorithms); high performance over a determined set of problems is compensated by medium and low performance in all other problems [1, 2]. Therefore, taking into account all possible problems, the overall performance for all possible optimization algorithms tends to be the same [1, 6–8].

Evolutionary algorithm behavior is governed by interrelated aspects of exploitation [1, 2]. Such aspects point to limitations that could be overcome by the use of hybrid evolutionary methods dedicated to optimizing the performance of direct and classical evolutionary approaches [1–5]. The hybridization of evolutionary algorithms has become relatively widespread due to their ability to deal with a considerable amount of real world complex issues usually constrained by some degrees of uncertainty, imprecision, interference, and noise [4, 5, 9, 10].

Academia and industry have been paying increasing interest to non-invasive imaging techniques and health applications [11, 12], since imaging diagnosis techniques and devices based on ionizing radiation methods could be related to the occurrence of several health problems due to long exposure, like benign and malignant tissues and, consequently, cancer, one of the most important public health problems, both for developed and under-developed countries [11, 12].

Electrical impedance tomography (EIT) is a low-cost, portable, and safely handled non-invasive imaging technique free of ionizing radiation, offering a considerably wide field of possibilities [13]. Its fundamentals are based on the application of electrical currents to a pair of electrodes on the surface of the volume of interest [13–16], returning electrical potentials used in tomographic image reconstruction, that is finding the distribution of electrical conductivities, by solving the boundary value problem [15, 16]. Since this is an ill-posed problem, there is no unique solution, that is there is no warranty to obtain the same conductivity distribution for a given distribution of electrical potentials on surface electrodes [13, 16].

Boundary value problems can be solved using optimization problems by considering one or more target metrics to optimize. Taking into account this principle, the EIT reconstruction problem can be solved by the effort to minimize the relative reconstruction error using evolutionary computation, where individuals (solution candidates) are probable conductivity distributions. The reconstruction error is defined as the error between the experimental and calculated distributions of surface voltages.

In this chapter we propose a methodology to solve the problem of reconstruction of EIT images based on hybrid evolutionary optimization methods in which tuning limitations are compensated by the use of adequate heuristics. We perform simulations and compare experimental results with ground-truth images considering the relative squared error. The evaluation of quantitative and qualitative results indicate that the use of hybrid evolutionary and bioinspired algorithms aid the avoidance of local minima and obtain anatomically consistent results, side-stepping the use of empirical constraints as in common and non-hybrid EIT reconstruction methods [17], in a direct and relatively simple way, despite their inherent complexity.

This chapter is organized as follows. In Sect. 2 we present a review on evolutionary computation and bioinspired algorithms, with special focus on swarm intelligence; in Sect. 3 we present a review on density-based fish school search, a bioinspired swarm algorithm based on fish school behavior; in Sect. 4 we briefly present a Gauss–Newton electrical impedance tomography reconstruction method; some comments on hybridization are presented in Sect. 6; in Sect. 5 we present a bibliographical revision of EIT, image reconstruction problems, and software tools for image reconstruction based on finite elements; in Sect. 7 we present our proposed EIT image reconstruction methodology based on hybrid heuristic optimization algorithms, as well as the experimental approach and infrastructure; in Sect. 8 we present the experimental results and some discussion; since EIT is a relatively new imaging technique, we also present a hardware proposal in Sect. 9; finally, in Sect. 10 we make more general comments on methodology and results.

2 Heuristic Search, Evolutionary Computation, and Bioinspired Algorithms

Evolutionary computation is one of the main methodologies that compose computational intelligence. The algorithms of evolutionary computation (called evolutionary algorithms) were inspired by the evolution principles from genetics and elements of Darwin's Theory of Evolution, such as natural selection, reproduction, and mutation [18].

The main goal of evolutionary computing is to provide tools for building intelligent systems to model intelligent behavior [18]. Since evolutionary algorithms are non-expert iterative tools, that is they are not dedicated to a specific problem, having a general nature that makes possible their application to a relatively wide range of problems, they can be used in optimization, modeling, and simulation problems [19].

An evolutionary algorithm is a population-based stochastic algorithm. According to [19], the idea behind evolutionary algorithms is the same: given a population of individuals embedded in some resource-constrained environment, competition for such resources causes natural selection, where better adapted individuals succeed in surviving, reproducing and perpetuating its characteristics (survival of the fittest). As the generations go by, the fitness of the population increases toward the environment. The environment represents the problem itself, an individual a possible solution (also called a solution candidate), the fitness of an individual represents the quality of the solution to the problem in question, and generations represent the iterations of the algorithm [19, 20]. The way in which these algorithms solve problems is called trial-and-error (also known as *generate-and-test*) [19]. Following evolutionary concepts, possible solutions to the problem are generated and evaluated for the problem in question. In order to get even better solutions, some of them are chosen to be combined by generating new candidates for the solution.

Besides evolutionary algorithms, computational intelligence also uses bioin-spired algorithms. These two types of algorithms differ from each other by the inspiration or metaphor taken into account for its development. While evolutionary algorithms take into account the theories of genetics and evolution, the other one takes into account the behavior of living things in nature, such as the collective behavior of birds in searching for food. The brute-force proposal *generate-and-test* is the same for both algorithms; however, bioinspired algorithms do not have recombination, mutation, and selection operators. In spite of that, they have operators which simulate intelligent behavior.

Some examples of bioinspired algorithms are: Particle Swarm Optimization (PSO), based on the behavior of birds in searching for food [21]; the Bacterial Foraging Algorithm, inspired by the social foraging behavior of the bacterium Escherichia coli present in the human intestine [22]; the Search for Fish School (FSS) [23]; and Density based on Fish School Search (dFSS) [24], based on the collective behavior of fish in searching for food. Some of these types of algorithm are also based on insect behavior: Ant Colony Optimization (ACO) [25] and Artificial Bee Colony [26, 27]. In the following subsections some of these algorithms that were applied in the reconstruction of EIT images are presented.

2.1 Particle Swarm Optimization

PSO is a bioinspired method for search and optimization inspired by a flock's flight in search of food sources and the shared knowledge among the flock's members. The method was created by mathematically modeling the birds as particles, their flight as direction vectors, and their velocity as continuous values [21].

As in the bird's flock flight, the particles move toward the food source influenced by two main forces: its own knowledge and memory of where to go, and the leader's experience. The leader is commonly the bird which is most experienced, and in the mathematical model it is the solution candidate that best fits the approached

function [21]. The velocity and weight distribution updating is done according to
the following expressions (1) and (2).

$$v_i(t+1) = \omega(t)v_i(t) + c_1(t)r_1(t)(p_i - x_i(t)) + c_2(t)r_2(t)(p_g - x_i(t)) \qquad (1)$$

$$x_i(t+1) = x_i + v_i(t+1) \qquad (2)$$

where $v_i(t+1)$ is the velocity vector of the i-th particle; $\omega(t)$ is the inertia factor;
c_1 and c_2 are the individual influence and the social factor, modeling the strength of
the particle's own experience and that of the leader, respectively; and $r_1(t), r_2(t) \sim U[0, 1]$.

Along the iterative process, the particles have their configuration updated
by receiving the mathematical influence of the best configuration found so far
(modeling the leader influence), and the best configuration ever found by the particle
itself (modeling its own experience). There is also an inertia factor, which is a
number, generally less than 1, that decreases the velocity of the particle at each
iteration, modeling the reduction of the velocity when the particle gets close to the
optimal configuration. In Table 1 we show the general PSO algorithm.

In the first step, the particles are initialized as vectors containing 'Dim' positions
each. Each position is normally initialized with a random number, in the range of
domain of the approached problem.

The second phase is to initialize the velocities. They are vectors with the same
dimension of the particles and in the same quantity of the particles. Each velocity
vector corresponds to a particle and each position of this vector corresponds to one
position (weight) of this particle.

The third and 5.III phases calculate the objective function for each particle and
create a rank with the evaluation of each particle. Those evaluations will be used in
the following step.

Table 1 Particle swarm optimization

Particle swarm optimization pseudo-code
1. Initialize the particles with random distribution
2. Initialize all the velocities as 0
3. Calculate the fitness function for all particles
4. Identify the best particle and the best position ever found for each particle
5. Until the number of iterations is reached, do:
(I)   Calculate the objective function for each particle.
(II) Update the velocities and weights of each particle.
(III) Identify the Pbest and GBest.
end Until

In steps 4 and 5.II, GBest is identified, which is the best particle found so far (by evaluation through the objective function), and Pbest is identified, which for each particle is the best configuration found so far.

Finally in the 5.II phase, the particle weights are updated by their respective velocity vector and are calculated as in the expression (1).

The PSO algorithm, as approached in this work, is used to reconstruct EIT images by modeling solution candidates as particles, as described in Sect. 7.

The hybrid technique will evolve the insertion of one of the particles that will be generated by the Gauss–Newton algorithm.

2.2 Simulated Annealing

Simulated annealing (SA) is a stochastic approach for the global optimization of a given function based on local search and hill climbing. It is a metaheuristic optimization algorithm to deal with large discrete search spaces [28]. SA could be preferable to gradient-descent methods when reaching an approximate global optimum is more important than finding a precise local optimum, given a determined time interval [28].

Its metaheuristic is inspired by annealing in metallurgy, in which heating and controlled cooling of a material is employed to increase the size of its crystals and reduce their defects, affecting both the temperature and the thermodynamic free energy. The process of slow cooling is interpreted as a slow decrease in the probability of accepting the worse solutions, thus improving exploration capabilities [28]. In each iteration, a solution close to the current one is selected and evaluated. Afterwards, the algorithm decides to accept or reject the current solution, taking into account the probabilities of the newly calculated solution being better or worse than the present solution. The temperature is progressively decreased, converging from an initial positive value to zero, which affects the probabilities. Probabilities are calculated as described in Eq. (3). The general behavior of SA algorithms is given by the pseudocode of Table 2, adapted to solve the EIT image reconstruction problem [29]. The symbolic function *GenerateRandomNeighbor(S)* is responsible for obtaining a neighbor solution in the search space of the current solution (S) and depends on the problem to be solved.

$$P(\Delta E) = e^{-\frac{\Delta E}{kT}} \tag{3}$$

where $P(\Delta E)$ is the probability of keeping a state that produces a thermal energy increase of ΔE (as in statistical mechanics) in the objective function; k is a parameter analogous to Stefan–Boltzman's Constant, usually assumed to be 1; and T is the temperature defined by a cooling scheme, the main parameter of the process control. The probability of a particular state decreases with its energy as temperature increases. This fact can be observed in the reduction of the slope of

Table 2 The pseudocode of a Simulated Annealing algorithm

Pseudocode: Simulated Annealing		
1.	$S* \leftarrow S$;	//Initial solution
2.	$T \leftarrow T_0$;	//Initial temperature
3.	k;	//Process parameter
4.	γ;	//Random constant $\in [0, 1]$
5.	$IterT_{max}$;	//Maximum iteration for temperature stabilization
6.	while Stopping criteria is not satisfied	
7.	$IterT \leftarrow 0$;	//Iterations for temperature stabilization
8.	while $(IterT < IterT_{max})$ do	
9.	$IterT \leftarrow IterT + 1$	
10.	$S' \leftarrow GenerateRandomNeighbor(S)$;	
11.	$\Delta E \leftarrow f(S') - f(S)$;	//Change of objective function f
12.	if $(\Delta E < 0)$ then	
13.	$S \leftarrow S'$;	
14.	if $(f(S') < f(S))$ then	
15.	$S* \leftarrow S'$;	
16.	else if	
17.	take a random $P \in [0, 1]$;	
18.	if $\left(P < e^{-\frac{\Delta E}{kT}}\right)$ then $S \leftarrow S'$;	
19.	end of if;	
20.	Temporary end;	
21.	$T \leftarrow \frac{T}{1+\gamma\sqrt{T}}$;	
22.	$IterT \leftarrow 0$;	
23.	Temporary end;	
24.	$S \leftarrow S*$;	
25.	Return $S*$;	
26.	End of Simulated Annealing	

$P(\Delta E)$ [29]. Additionally, it is possible to demonstrate that, as SA converges to the global minimum, a very slow temperature reduction is observed, requiring a considerably large amount of iterations [28].

2.3 Differential Evolution

After presenting the Chebychev Polynomial Problem by Rainer Storn, Kenneth Price (1995) in attempting to solve this problem created Differential Evolution (DE) when he had the idea of using different vectors to recreate the population vector. Since then, a number of tests and substantial improvements have been made which has resulted in DE becoming a versatile and robust Evolutionary Computation algorithm [30, 31].

Table 3 The pseudocode of a Differential Evolution algorithm

Pseudocode: Differential Evolution

1. Generate the initial population of n random agents, each one represented by a vector x_i where $i = 1, 2, \ldots, n$
2. Repeat until the stop criterion is satisfied:
   (A) For $i = 1, 2, \ldots, n$ do

      (i) Given a random number $r \sim U[0, 1]$;
     (ii) If $r \leq P_{CR}$ then

         (a) Generate a mutated agent through the Table 4;
         (b) Generate a cross-agent $\check{x}_i$ through the expression (4);
         (c) If $f_0(\check{x}_i) < f_0(x_i)$ then $x_i \leftarrow \check{x}_i$ (minimization);

2.3.1 Stages of Differential Evolution

According to [30], DE is characterized by its simplicity and efficiency at solving global optimization problems in continuous spaces. Similar to Genetic algorithms, DE benefits from diversity combined operators of mutation and crossover, in order to generate individuals modeled as vectors, which are candidates for the next generation. The new population is defined by the selection mechanism, selecting the individuals to survive for the next population according to simple criteria. In this process, the population size remains constant. Therefore, the individuals of generation G are modeled as real vectors, $x_{i,G}$, $i = 1, 2, \ldots, N_P$, where N_P is the population size [32].

The optimization process is governed by the following:

1. **Initialization:** The algorithm is initiated by creating a randomly chosen initial population with uniform distribution, corresponding to the vectors of unknown parameters (potential solutions), taking into account the search space limit [30]. Typically, the unknown parameters are conditioned by lower and upper boundary constraints, $x_j^{(L)}$ e $x_j^{(U)}$, respectively, as in the following:

$$x_j^{(L)} \leq x_j \leq x_j^{(U)}, j = 1, 2, \ldots, D. \tag{4}$$

   Therefore, the initial population is defined as:

$$x_{j,i}^{(0)} = x_j^{(L)} + r_j[0, 1] \cdot (x_j^{(U)} - x_j^{(L)}) \tag{5}$$

   where, $i = 1, 2, \ldots, N_P$, $j = 1, 2, \ldots, D$, and $r_j \sim U[0, 1]$.

2. **Mutation:** In DE, new individuals are generated by three individuals, related as following. For each individual $x_{i,G}$, $i = 1, 2, \ldots, N_P$, a basic DE mutant vector (classical) is generated according to:

$$v_{i,G+1} = x_{a,G} + F(x_{b,G} - x_{c,G}), \tag{6}$$

Table 4 Main strategies of the Differential Evolution technique

Mutation expression	Notation
$V^{q+1} = X_\alpha^q + F_P(X_\beta^q - X_\gamma^q)$	ED/rand/1/bin
$V^{q+1} = X_\alpha^q + F_P(X_\beta^q - X_\beta^q) + F_P(X_\gamma^q - X_\delta^q)$	ED/rand/2/bin
$V^{q+1} = X_{best}^q + F_P(X_\beta^q - X_\gamma^q)$	ED/best/1/bin
$V^{q+1} = X_{best}^q + F_P(X_\alpha^q - X_\beta^q) + F_P(X_\gamma^q - X_\delta^q)$	ED/best/2/bin
$V^{q+1} = X_{old}^q + F_P(X_{best}^q - X_{old}^q) + F_P(X_\gamma^q - X_\delta^q)$	ED/rand-to-best/2/bin
$V^{q+1} = X_\alpha^q + F_P(X_\beta^q - X_\gamma^q)$	ED/rand/1/exp
$V^{q+1} = X_{best}^q + F_P(X_\beta^q - X_\gamma^q)$	ED/best/1/exp
$V^{q+1} = X_\alpha^q + F_P(X_\rho^q - X_\beta^q) + F_P(X_\gamma^q - X_\delta^q)$	ED/rand/2/exp
$V^{q+1} = X_{best}^q + F_P(X_\alpha^q - X_\beta^q) + F_P(X_\gamma^q - X_\delta^q)$	ED/best/2/exp
$V^{q+1} = X_{old}^q + F_P(X_{best}^q - X_{old}^q) + F_P(X_\gamma^q - X_\delta^q)$	ED/rand-to-best/2/exp

where the random indexes $a, b, c \in \{1, 2, \ldots, N_P\}$ and $a \neq b \neq c$. The amplification parameter $F \sim U(0, 2]$ controls the amplification of the differential variation $(x_{b,G} - xc, G)$. However, there are some other variant mutation operations, as shown in Table 4.

A potential reason for DE to acquire reasonable results is that the mutation operator is governed by the difference between the coordinates of the individuals of the current population [30]. Consequently, each parameter is automatically exchanged and appropriately reduced, aiding convergence to the desired approximate solution.

3. **Crossover:** The mutated vector is mixed with the target vector to produce the trial vector, formed as in the following:

$$w_{i,G+1} = \begin{cases} v_{i,G+1}, \ r_j \leq CR \quad \vee \quad j = k(i) \\ x_{i,G}, \quad r_j > CR \quad \vee \quad j \neq k(i) \end{cases}, \tag{7}$$

where $j = 1, 2, \ldots, D$, $i = 1, 2, \ldots, N_P$, $r_j \sim U[0, 1]$, and $k(i) \in 1, 2, \ldots, D$ is a randomly chosen index, which ensures that $w_{i,G+1}$ receives at least one coordinate of $v_{i,G+1}$. The parameter $CR \in [0, 1]$ is the crossing constant, set by the user [30].

4. **Selection:** The individual of the new generation is selected as the better evaluated vector from the trial vector $w_{i,G+1}$ and the target vector $x_{i,G}$, according to the objective function [30].

5. **Stop criterion:** The iterative process finishes when a determined number of iterations is reached or a predetermined value of the objective function is obtained with a considerably small error [30].

6. **Selection of control parameters:** The population size N_P is chosen from $5D$ to $10D$ and is kept constant during the search process. Parameters F and CR are set during the search process and affect the speed of convergence and robustness. Appropriate values for NP, F, and CR are usually empirically determined [30].

The optimization process is detailed in the pseudocode of DE algorithms, designed to minimize an objective function $f_0 : \mathbb{R}^n \to \mathbb{R}$, where P_{CR} is the probability of crossing [33–36], shown in Table 3.

2.3.2 Differential Evolution Strategies

The strategies of DE consist of different variation operators, and can be nominated according to the following acronym: DE/a/b/c, where [32]:

a: Specifies the vector to be disturbed, which can be "rand" (a vector of the randomly selected population) or "best" (the least cost vector of the population);
b: Determines the number of weighted differences used for the perturbation of a;
c: Denotes the type of crossings (exp: exponential; bin: binomial).

Price [37] suggested 10 different strategies and some work guidelines for using them. These strategies were derived from the five different DE mutation regimes. Each mutation strategy was combined with the "Exponential" or "Binomial" crossover, providing $5 \times 2 = 10$ DE strategies. Nevertheless, other combinations of linear vectors can be used for the mutation. In Table 4 we have Prive's 10 DE strategies [37].

The α, β, γ, ρ, and δ indexes are mutually exclusive integers chosen randomly in the range $[1, n]$, where n is the number of agents of the initial population, and all are different from the *old* base index. These indexes are randomly generated once for each donor vector. The scale factor F_p is a positive control parameter for the expansion of the difference vectors. $X_{best}^{(q)}$ is the agent vector with the best aptitude (i.e. the lowest objective function value for a minimization problem) in the generation population q.

2.4 Fish School Search

The FSS was developed by Bastos Filho and Lima Neto in 2008. Such a method takes into account the protection and realization of the mutual achievements of oceanic fish [23], where the real characteristics of the fish that served as inspiration for the method can be classified as feeding and swimming. In the first characteristic, it is considered that fish possess the natural instinct to seek food; such a characteristic is incorporated into the algorithm to indicate the success of fish (those who find more feed are more successful). In the second characteristic, the capacity of the fish to move individually and in a flock (shoal) is taken into account, where the search for feed is one of the reasons why fish move.

Therefore, the process of FSS is performed by a population of individuals with limited memory—the fish [23]. Each fish in the shoal represents a possible solution to the optimization problem [23]. This method is recommended for high-dimension search and optimization problems [23, 24]. Considering the characteristics of feeding and swimming, the algorithm is formed by four operators, being one of feeding (the weight) and three of swimming (the operators of movement), which are described in the following sections.

2.4.1 Individual Movement Operator

Considering fish's individual capacity for searching for food, the individual movement operator is responsible for moving the fish to a random region of its neighborhood; this movement does not consider the influence of the other fish of the school for its realization. An important feature of such a movement is that the fish moves only in the positive direction of the fitness function, that is it travels only if the random position is better than the current position.

The individual displacement of each fish i, Δx_{ind_i} is given in Eq. (8), where $rand(-1, 1)$ is a vector of random values uniformly distributed in the range $[-1, 1]$, and $step_{ind}$ is the individual movement step, a parameter that represents the ability of the fish to move in the individual movement. After calculating the individual displacement the position of fish x_{ind_i} is updated through Eq. (9).

$$\Delta x_{ind_i}(t + 1) = step_{ind} \cdot rand(-1, 1) \tag{8}$$

$$x_{ind_i}(t + 1) = x_{ind_i}(t) + \Delta x_{ind_i}(t + 1) \tag{9}$$

In order to guarantee the convergence of the algorithm during the search process, the value of the parameter $step_{ind}$ decreases linearly as shown in Eq. (10), where $step_{ind_i}$ and $step_{ind_f}$ are the initial and final values of $step_{ind}$, and $iterations$ is the maximum possible value of iterations of the algorithm.

$$step_{ind}(t + 1) = step_{ind}(t) - \frac{step_{ind_i} - step_{ind_f}}{iterations} \tag{10}$$

2.4.2 Feeding Operator

The weight of the fish is the indicator of its success, that is the more food the fish finds, the more successful the fish is, which represents a better solution in the optimization problem [24, 38]. In this way, weight is the function to be maximized by the search process. The fish's weight is given as a function of the variation of the fitness function generated by the individual movement (Δf_i), as shown in the equation below:

$$W_i(t + 1) = W_i(t) + \frac{\Delta f_i}{\max(\Delta f)} \tag{11}$$

where $W_i(t)$ and $W_i(t+1)$ represent the weight of the i fish before and after the update.

2.4.3 Collective Instinctive Movement Operator

The first collective movement to be considered in FSS is the collective-instinctive movement, where the most successful fish in the individual movement guides the other fish to the points of greatest food encountered by it. Such motion is carried out through the resulting direction vector $I(t)$, that is the weighted average of the individual displacements ξ. Having as weight the individual fitness variance Δx_i, the expression for that mean is given in Eq. (12), where N represents the number of fish in the shoal. Then, the new position of all fish is obtained following Eq. (13).

$$I(t) = \frac{\sum_{i=1}^{N} \Delta x_{ind_i} \Delta f_i}{\sum_{i=1}^{N} \Delta f_i} \tag{12}$$

$$x_i(t+1) = x_i(t) + I(t) \tag{13}$$

2.4.4 Collective Volitive Movement Operator

The following and last collective movement, the collective-volitional movement, is based on the performance of all fish in the school [39]. In this movement, the fish may move towards the center of mass of the shoal or move away from it. The calculation of the center of mass of the school, $Bary(t)$, is done according to Eq. (14).

The choice whether the fish will approach or move away from the center of mass is made by analyzing whether the fish are gaining weight during the search process. If the fish, in general, are increasing in weight, it means that the search is being successful and the fish approach each other, decreasing the radius of the search; in this case, the movement is carried out following Eq. (15). Otherwise, if the fish are losing weight, the search is unsuccessful and the fish move away from each other, increasing the search radius, executing the movement through Eq. (16). In Eqs. (15) and (16), the parameters $rand(0, 1)$ represent a vector of random values evenly distributed in the range $[0, 1]$, and $step_{vol}$ is the step of the collective-volitional movement.

$$Bary(t) = \frac{\sum_{i=1}^{N} x_i W_i(t)}{\sum_{i=1}^{N} W_i(t)} \tag{14}$$

$$x(t+1) = x(t) - step_{vol} \cdot rand(0, 1)(x(t) - Bary(t)) \tag{15}$$

$$x(t+1) = x(t) + step_{vol} \cdot rand(0, 1)(x(t) - Bary(t)) \tag{16}$$

3 Density Based on Fish School Search

Based on the FSS, the Density based on Fish School Search (dFSS) is an algorithm dedicated to the optimization of multimodal functions proposed by Madeiro, Bastos-Filho, and Lima Neto [24, 38]. In the method, the main Fish School is divided into Sub-Fish Schools of different sizes, so that each sub-group will explore different regions of the aquarium, which have possible solutions to the problem. Unlike the other methods discussed here, which get only one solution at the end of their runs, on dFSS it is possible to obtain a set of global and local optimal solutions. In fact, the function of the relative quadratic error for TIE is multimodal, but since the objective of the reconstruction is to obtain only one image, the application of dFSS to TIE is done by considering as a solution the best image obtained by the method in relation to the objective function.

In addition to the feeding and movement operators of the search for Fish schools adapted to the multimodal approach, dFSS has two more operators, the memory and partitioning operators, which will be dealt with in more detail later in this section.

In dFSS the food purchased by a fish is shared with the other fish of the shoal. The amount of shared food from one fish i to another fish j, $C(i, j)$ is given by Eq. (17), where q_{ij} is the number of fish k that satisfy the relationship $d_{ik} < d_{ij}$ (density of fish around the fish i), including fish i, and $d_{R_{ij}} = d_{ij}/[\forall k \neq i, \min(d_{ik})]$ is the normalized distance. Then, the updating of the weight of each fish will take into account the total food that was shared with it, as given by Eq. (18), where Q represents the number of fish that were successful during the individual movement. Different from what happens in nature, in the dFSS proposal it is assumed that the weight of the fish does not decrease over the iterations [24].

$$C(i, j) = \frac{\Delta f_i}{(d_{R_{ij}})^{q_{ij}} \sum_{k=1}^{N} \frac{1}{(d_{R_{ik}})^{q_{ik}}}} \tag{17}$$

$$W_i(t + 1) = W_i(t) + \sum_{j=1}^{Q} C(i, j) \tag{18}$$

Food sharing is responsible for the control and maintenance of different sub-Fish Schools, since each fish cooperates (sharing its success) with the other fish around them, the most significant sharing is with the closest fish and in regions less populous. This is modeled by the index $(d_{R_{ij}})^{-q_{ij}}$ greater is the value of q_{ij} (i.e. the denser the region around the fish i), smaller will be the quantity shared for the fish j [24]. Further detail on the formation of sub-shoals is given in the explanation of the memory and partitioning operators in the following paragraphs.

Individual movement in the dFSS occurs in the same way as in the FSS, but for segregation of the main fish school the adjustment of the movement parameter has been modified. The new way of updating the step of the individual movement is given by Eqs. (19), (20), (21), and (22), where $decay_i$ is the decay rate, $decay_{\min} \in [0, 1]$, $decay_{\max}$ are the minimum and maximum decay rates, respectively, $decay_{\max_{init}}$, $decay_{\max_{end}} \in [0, 1]$ are the initial and final maximum decay rates, respectively, and must obey the following condition: $decay_{\max_{end}} < decay_{\max_{init}} < decay_{\min}$, and finally $T_{\max}$ is the maximum number of iterations [24, 38]. The initial value of the individual step is given by the parameter $step_{init}$, that is $step_{ind_i}(0) = step_{init}$ [24].

$$step_{ind_i}(t + 1) = decay_i \cdot step_{ind_i}(t) \tag{19}$$

$$decay_i = decay_{\min} - \left(\frac{R_i(t) - \min(R_j(t))}{\max(R_j(t)) - \min(R_j(t))} \right) (decay_{\min} - decay_{\max}(t)) \tag{20}$$

$$decay_{\max}(t) = decay_{\max_{init}} \left(\frac{decay_{\max_{end}}}{decay_{\max_{init}}} \right)^{t/T_{\max}} \tag{21}$$

$$R_i(t) = \sum_{j=1}^{Q} \frac{\Delta f_j}{(d_{R_{ij}})^{q_{ij}} \sum_{k=1}^{N} \frac{1}{\left(d_{R_{jk}}\right)^{q_{jk}}}} \tag{22}$$

In the dFSS each fish has a memory $M_i = \{M_{i1}, M_{i2}, \ldots, M_{iN}\}$, where N is the total number of fish. In a fish's memory is the information of how much food the other fish have shared with it throughout the search process. The index M_{ij} indicates the influence of fish j on fish i, that is, the larger the M_{ij} the greater the influence of fish j on fish i. The memory operator is calculated by Eq. (23), where $\rho \in [0, 1]$ is the forgetting rate, a parameter that controls how the influence exerted on past iterations is remembered.

$$M_{ij}(t + 1) = (1 - \rho)M_{ij}(t) + C(j, i) \tag{23}$$

The collective-instinctive movement of dFSS is similar to the Fish School search, although each fish has its own resulting direction vector I_i which is given by the weighted average of the displacement performed on the individual movement by a fish j having as weight its influence M_{ij} as shown in Eq. (24). The fish position update is given by Eq. (25). According to [24], even the fish that did not move during the individual movement will influence the result of I_i, causing fish i to simulate this behavior by remaining stationary according to the value of M_{ij}.

$$I_i(t) = \frac{\sum_{j=1}^{N} \Delta x_j M_{ij}}{\sum_{k=1}^{N} M_{ik}} \tag{24}$$

$$x_i(t + 1) = x_i(t) + I_i(t) \tag{25}$$

After the execution of the collective-instinctive movement the partitioning operator responsible for the division of the main school into several sub-schools of different sizes is executed. For the division of the school, the following belonging condition is taken into account: a fish i belongs to the same sub-school of fish j only if i is the fish that exerts the greatest influence on fish j or vice versa [24]. The division process begins when a randomly chosen fish i is removed from the main school to form a new sub-shoal, then another fish j is sought where i is the most influential for j or vice versa; if there is a j that satisfies this condition it will be removed from the main shoal and added to the sub-school in question; then this procedure is repeated for fish j in a cascade process. This process is repeated until no more fish that satisfy the condition of belonging to that particular sub-school are found. When this happens a new fish from the main shoal will be randomly removed to compose a new sub-school and the process resumes. The formation of the sub-shoals is carried out until there are no more fish in the main shoal [24].

Finally, the collective-volitional movement is performed independently for each sub-school as given in Eq. (26). In this movement all the fish move towards the barycenter of the sub-shoal $(Bary_k(t))$ to which it belongs. The barycenter of each sub-school is calculated in the same way as FSS, given in Eq. (14). To avoid premature convergence, the magnitude of the pitch to be performed by the fish towards the barycenter varies according to the value of $decay_{\max}(t)$ [24]. The pseudocode of the dFSS method [38] is shown in Table 5.

$$x(t + 1) = x(t) + (1 - decay_{\max}(t))(Bary_k(t) - x(t)) \tag{26}$$

4   The Gauss–Newton Method

Based on Newton's method (dedicated to estimating roots of a function) the Gauss–Newton method is an algorithm which has been widely used in the reconstruction of EIT images [40, 41]. This method, implemented to eliminate the use of second derivatives, consists of a gradient-descent-based numerical method used to solve non-linear least squares problems minimizing the sum of quadratic functions [41]. The application of the Gauss–Newton method in EIT is done by estimating a conductivity distribution σ_k which minimizes the expression given in (27), where $\phi_{ext,k}(\overrightarrow{u}) = f(I(\overrightarrow{u}), \sigma_k(\overrightarrow{v}))$, for all $\overrightarrow{u} \in \partial\Omega$ and $\overrightarrow{v} \in \Omega$ [41].

$$s_k = \frac{1}{2} \sum_{u \in \partial\Omega} (\phi_{ext,k}(\overrightarrow{u}) - \phi_{ext}(\overrightarrow{u}))^2 \tag{27}$$

The hybridizations done in this work evolve the Gauss–Newton method, working together with the techniques presented before. The nomenclature used to describe a hybrid technique in this works is 'technique' with Non Blind Search (NBS) or 'technique'-NBS. For instance, the hybrid approach with the PSO and the Gauss–Newton algorithm will be called a PSO a with non-blind search, or PSO-NBS.

Table 5 The pseudocode of density based on Fish School Search algorithm

Pseudocode: density based on Fish School Search

1. Initialize all fish in random positions and with null weight
2. Evaluate all fish's fitness
3. Calculate the distance between the fish
4. Repeat steps (a) and (b) until some stopping criteria are reached

   (a) For each fish in the school do:

      (I) Execute the individual movement
      (II) Execute the feed operator
      (III) Update the fish's weight
      (IV) Execute the memory operator

   (b) For each fish in the school do:

      (I) Execute the instinctive collective movement
      (II) Determine the most influential fish for a given fish

   (c) Execute the primary school division operator
   (d) For each sub-school determined in previous step do:

      (I) Calculate the barycenter

   (e) For each fish do:

      (I) Update the individual step
      (II) Execute the collective-volitional movement

   (f) Evaluate all fish's fitness
   (g) Calculate the distance between the fish
   (h) Update the value of $decay_{max}(t)$

5. Select the best fish from each sub-school at the end of the process

---

5 Electrical Impedance Tomography

EIT is a promising imaging technique that is non-invasive and free of ionizing radiations. This technique reconstructs images of the inside of a body (or any object) through electrical quantities measured below its surface. For this, electrodes are placed across a transverse section of a body. These electrodes are connected with a control and data acquisition system that is responsible for the application of a pattern of an alternated and low amplitude electrical current and also for the measurement of the border electrical potentials that are generated by this stimulus. When collected, the electrical current and potential data are conveyed to a computer that makes the image reconstruction [42, 43].

The images obtained by EIT are the computational reproduction of an estimated mapping of electrical properties inside a section of body that are calculated through the relation between the stimulus data and response data. By electrical properties, in this case, is understood electrical conductivity or permissiveness; electrical conductivity is the measurement of how easy a material conducts electricity and electrical permissiveness is the measurement of how easy electrical charges of a material are separated under the application of an electric field. A good conductor allows passage of continuous and alternated current whereas a high permissiveness material allows the passage only of alternated current [44]. The tissues and organs of the human body, due to its constitution, have characteristic values of conductivity and permissiveness. Factors such as concentration of water, ions, and blood irrigation determine whether a tissue or an organ is less or more conductive or permissive. This justifies the use of EIT in the medical field, because the difference of conductivity and permissiveness between organs supplies the necessary contrast in the EIT image for differentiation of these organs.

Nowadays, the medical imaging field has very well consolidated techniques such as X-rays, computed tomography, nuclear magnetic resonance, and positron emission tomography. Despite that, EIT has advantages when compared to these other methods that makes it a promising technique in medical imaging. These advantages are:

- EIT does not use ionizing radiation, hence is harmless to patients [44]. Thus, it is viable for performing this technique several times or even for continuous use in monitoring certain functionalities of a body.
- EIT has small dimensions [45], which allows the device to be moved to the patient and even used in the patient's bed in intensive care units.
- EIT has low cost when compared to other imaging techniques.

Although these advantages make EIT a promising technique, it is still recent and not strongly established, presenting low-resolution images and slow reconstruction when compared to other tomography techniques [46, 47], which makes it not fully reliable for medical diagnosis.

The process of reconstructing EIT images is divided into two problems: the direct problem and the inverse problem [46, 47]. In the direct problem, the conductivity distribution of the inner domain and the current are known and the objective is to determine an electric potential distribution at the internal and boundary (edge) points of the domain [45]. This process is governed by Poisson's equation and its boundary conditions for the EIT problem. In the inverse problem, the goal is to estimate the internal conductivity distribution of a domain knowing the pattern of the current excitation and the edge potentials due to this excitation. The inverse problem, mathematically, is a non-linear, poorly placed and poorly conditioned problem [42, 48], because more than one solution is possible for this problem for the same input values, that is small data measurement errors can lead indefinitely to large errors in the solution [42, 48]. This factor makes EIT imaging quite dependent on the reconstruction method used.

From a mathematical point of view, the field of study of EIT, that is the section for which the image is desired, can be considered to be a closed 2D region Ω where its boundary surface is given by $\partial \Omega$ as shown in Fig. 1.

Fig. 1 EIT domain

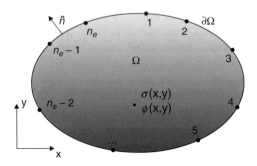

To reduce the complexity of this problem it is considered that the image domain should consist of an isotropic medium.[1] However, the electrical nature of human organs and tissues is anisotropic[2] [49, 50]. Despite the proposed hypothesis being wrong, it is necessary, due to the limited knowledge about this topic in EIT and related areas [49].

Taking into account a low-frequency excitation current (of the order of 125 kHz) the permissiveness effect can be disregarded [51]. Thus, the electric medium is considered to be conductive $\sigma(x, y)$ because, when considering low frequencies, the inductive and capacitive effects can be ignored [52]. Therefore, the current density $\vec{J}$ generated from an injected electric current is given by Eq. (28) [51].

$$\vec{J} = \sigma \vec{E} \tag{28}$$

where $\vec{E}$ represents the medium's electric field. Considering that the excitation frequency value is lower than 30 MHz, it has the following [53]:

$$\vec{E} = -\nabla \phi \tag{29}$$

$$\nabla \cdot \vec{J} = 0 \tag{30}$$

where ∇ is the symbol *nabla* that denotes the gradient operator,[3] $\nabla\cdot$ is the divergent operator[4] and $\phi(x, y)$ represents the internal electric potential at a point (x, y) of the domain Ω. Thus, replacing Eqs. (29) and (30) in Eq. (28) yields Poisson's equation, given in (31), which relates the conductivity values and electrical potentials of a domain [51, 53].

$$\nabla \cdot (\sigma \nabla \phi) = 0 \tag{31}$$

---

[1] An isotropic material is a medium whose electrical characteristics do not depend on the considered direction.

[2] Instead of isotropic materials, an anisotropic material has direction-dependent characteristics.

[3] The gradient of $F(x, y) = \nabla F(x, y) = i \frac{\partial F(x,y)}{\partial x} + j \frac{\partial F(x,y)}{\partial y} = \left(\frac{\partial F(x,y)}{\partial x}, \frac{\partial F(x,y)}{\partial y} \right)$.

[4] The divergent of $F(F_x, F_y) = \nabla \cdot F(F_x, F_y) = \frac{\partial F_x}{\partial x} + \frac{\partial F_y}{\partial y}$.

Poisson's equation has unlimited solutions, which means that for a given electrical potential distribution there are several conductivity distributions that satisfy Eq. (31). The number of solutions is limited by boundary conditions inherent to the problem. In EIT, the electrical currents are injected only by electrodes placed around the patient, which means that in specific positions on the domain's surface the following boundary condition can be taken into account:

$$\sigma \frac{\partial \phi(\sigma)}{\partial \hat{n}} = \begin{cases} J_i, \ i = 1, 2, 3, \ldots, n_e \\ 0, \ \text{in other points of } \partial \Omega \end{cases} \tag{32}$$

where n_e is the number of electrodes used and $\hat{n}$ is a normal versor[5] on the domain's edge and outside oriented. Following the same line, the known electrical potentials are the ones arranged on a domain's contour, measured by the electrodes. In this way, the second contour condition for this problem is the following:

$$\phi_{ext}(x, y) = \phi(x, y), \quad \forall (x, y) \in \partial \Omega \tag{33}$$

where $\phi_{ext}(x, y)$ is the electrical potential distribution measured by the electrodes.

Determination of the electric potential distribution measured by electrodes $\phi_{ext}(u, v)$, knowing the electric current of excitation $I(u, v)$, and distribution of internal conductivity $\sigma(x, y)$, is called the direct problem of EIT. It is defined by Eq. (31) and the boundary conditions (32) and (33) [51]. The direct problem can be modeled by the relation given in Eq. (34).

$$\phi_{ext}(u, v) = f(I(u, v), \sigma(x, y)), \quad \forall (u, v) \in \partial \Omega \wedge (x, y) \in \Omega \tag{34}$$

The inverse problem, however, is the reconstruction of EIT images themselves [42]. The objective is to determine an internal conductivity distribution $\sigma(x, y)$ in the domain by knowing the excitation current $I(u, v)$ and the edge potentials measured at the electrodes $\phi_{ext}(u, v)$. This problem is considered the inverse of the function given in Eq. (34), being modeled conformed in Eq. (35).

$$\sigma(x, y) = f^{-1}(I(u, v), \phi_{ext}(u, v)), \quad \forall (u, v) \in \partial \Omega \wedge (x, y) \in \Omega \tag{35}$$

Then, the direct problem is to solve Poisson's equation (31) knowing the internal conductivity distribution of a domain and the boundary condition given in Eq. (32) for the injected current. The inverse problem consists of the resolution of Eq. (31) knowing the two boundary conditions given by Eqs. (32) and (33), but not knowing the conductivity distribution [54].

The inverse problem of EIT is an intrinsically ill-posed problem because it does not have a unique solution, that is several conductivity distributions would respond to the current excitation at the same distribution of measured electrical

---

[5]A versor is a vector of unitary module usually used to indicate the direction in a given operation.

potentials. According to [49], if measurements were made with infinite precision and over the entire surface of the domain, the problem would have a unique solution. However, in the imaging process the data are discretely sampled and noisy, causing a loss of information. Besides that a large variation of conductivity can produce only a small variation in discrete measurements. Thus, the ideal would be to use as many electrodes as possible. It was found that by increasing the number of electrodes it is possible to improve the ill-posed condition of this problem and consequently the quality of the reconstructed images [55]. However, it has also been noted that increasing the number of electrodes significantly increases the reconstruction time [55]. In addition to that, the number of electrodes is limited by the measurement area and the size of the electrodes [56]. The inverse problem is also ill-conditioned because small oscillations in the measurements (such as noise) can produce large oscillations in the final non-linear solution, because changes in conductivity values of the domain do not produce a linear change in the values of the surface potentials [49].

5.1 Objective Function

The use of evolutionary and bioinspired algorithms in the reconstruction of EIT images occurs when approaching the reconstruction problem as an optimization problem. For this it is necessary that an objective function be optimized. In this chapter, is used as objective function the relative squared error given in Eq. (36), where n_e is the number of electrodes, V is the electrical potential distribution measured on the electrodes, and $U(x)$ is the electrical potential distribution of a random image x that is a candidate for the solution considered in the algorithm [57].

$$f_o(x) = \left[\frac{\sum_{i=1}^{n_e} (U_i(x) - V_i)^2}{\sum_{i=1}^{n_e} (V_i)^2} \right]^{1/2} \tag{36}$$

$$V = (V_1, V_2, \ldots, V_{n_e})^T, \tag{37}$$

$$U(x) = (U_1(x), U_2(x), \ldots, U_{n_e}(x))^T \tag{38}$$

5.2 Electrical Impedance Tomography and Diffuse Optical Tomography Reconstruction Software

Electrical Impedance Tomography and Diffuse Optical Tomography Reconstruction Software (EIDORS) is an open source software developed at MATLAB and Octave. The experiments presented in this chapter were performed using EIDORS, which has functions capable of solving the direct problem of EIT and the creation of finite element meshes[58].

6 Hybridization

Hybridization consists in using more than one technique in cooperation to solve a determinate problem, although if one technique can solve the approached problem in a satisfactory way, there is no necessity for hybridization. The use of hybridization is justified when the interaction among two or more techniques can improve the performance of the problem resolution.

There are mainly three types of hybrid system. They are:

- Sequential hybrid system: The technique is used in a pipeline way.
- Auxiliary hybrid system: These are techniques co-working to help one technique to solve a determinate problem. The accessory technique is used, normally, to improve the stages of the main technique.
- Embedded hybrid system: The evolved techniques are integrated. They work together as equals to approach the solution of a problem.

The hybridization described in this work are embedded, once the Gauss–Newton method is responsible for including a solution into the pool of solution candidates of each technique, which is intended to guide the search and to avoid falling in local minimals. The results are shown for EIT image reconstruction, by using raw techniques, such as particle swarm algorithms and density based on the Fish School Search; Also shown are the results of the collaborative work of each one with the Gauss–Newton model as new hybrid approaches.

6.1 Differential Evolution Hybridized with Simulated Annealing

According to [59] the DE method which has a stronger global search character, only a few generations are sufficient to find a solution close to the ideal. However, according to its algorithm, each generation requires the selection, crossing, and mutation of the agents of the current population, thus requiring a high computational cost, that is a high number of calculations of the objective function. The SA method is a local search algorithm with a high convergence speed due to the fact that it is able to avoid local minima [60, 61]. In this direction, the proposal of this work is a hybrid version of DE based on simulated annealing (DE-SA), which consists of the implementation of DE and adding SA within the selection operator to improve ED global search capacity.

Pseudocode

1. **Initialization:** Generate the initial population of n random agents D dimensional, each represented by a vector $X_{j,i,G} = rand_{j,i}[0, 1]$, where $j = 1, 2, \ldots, NP$; G is the current generation; and $F_i = rand(0, 2]$ is the mutation scale factor for each individual.
2. **Set the value of P_{CR};**
3. **Population Mutation:** The population mutation is based on the strategy of DE/best/1/bin. As shown in Table 4,

$$V_{j,i,G} = X_{j,best,G} + F_i(X_{j,i_1,G} - X_{j,i_2,G})$$

where $i_1 \neq i_2 \neq i$ and $X_{j,best,G}$ corresponds to the most suitable agent in the current generation.
4. **Population Crossing:** Population crossing is based on the DE binomial crossing operation, as shown by the equation of Table 4.

$$W_{j,i,G} = \begin{cases} V_{j,i,G}, & if \ (rand_{j,i}(0, 1) \ \leq \ P_{CR} \ or \ j = j_{rand}) \\ X_{ji,G}, & otherwise \end{cases}$$

5. **Population selection:** This is processed by comparing the target vector $X_{j,i,G}$ with the vector of judgment $W_{j,i,G}$ of the population. In addition, SA is added inside the selection operator, and t_G represents the ambient temperature of the current generation.

$$t_{G+1} = \frac{t_G}{1 + G\sqrt{t_G}} \tag{39}$$

$$x_{i,j,G+1} = \begin{cases} W_{j,i,G}, & if \ f_0(W_{j,i,G}) \leq f_0(W_{j,i,G}) \\ W_{j,i,G}, & if \ f_0(W_{j,i,G}) > f_0(W_{j,i,G}) \\ & and \ rest \left(\frac{G}{4}\right) = 0 \\ & and \ f_0(W_{j,best,G}) = f_0(W_{j,best,G-1}) \\ & and \ f_0(W_{j,best,G}) = f_0(W_{j,best,G-2}) \\ & and \ \exp\left[-\frac{(f_0(W_{j,i,G}) - f_0(W_{j,i,G}))}{t_G}\right] > rand(0, 1) \\ W_{j,i,G}, & Otherwise \end{cases} \tag{40}$$

where $W_{j,best,G}$ is the fittest agent of the current generation, $X_{j,best,G-1}$ is the fittest agent of generation $(G-1)$ and $X_{j,best,G-2}$ is the fittest agent of generation $(G - 2)$. Every four generations, these conditions are implemented to assess whether the value of the objective function of the best candidate for the solution has assumed a local minimum in the last three generations. In this case, the judgment vector $W_{j,i,G}$ can be maintained for the next generation if it satisfies the condition imposed by the exponential expression based on the probability of the SA optimization process.

6. **Stop if the stop criterion is satisfied. Otherwise, go back to Step 3.**

Recall that P_{CR} is the probability of crossing and f_0 corresponds to the objective function. Also for the DE-SA method, the initial agents were defined with a normal distribution of random conductivity in the range [0, 1].

6.2 Fish School Search Hybridization

According to the discussion in Sect. 2.4, one can observe that the FSS algorithm is quite dependent on the individual movement operator and therefore the parameter $step_{ind}$ decreases linearly with the iterative process.

With the proposal to increase the refined search between the intermediary and final iterations, we changed the linear decay for an exponential decay. In this way, the $step_{ind}$ value decays faster, thus the algorithm will execute a more refined search, that is the algorithm will execute a more exploitative and less exploratory search.

The equation for the exponential decay was made in such a way that the $step_{ind}$ value continues, beginning with $step_{ind_i}$ and ending with $step_{ind_f}$. This expression is given in Eq. (41).

$$step_{ind}(t) = step_{ind_i} \times \exp\left[\frac{t}{N} \ln\left(\frac{step_{ind_f}}{step_{ind_i}}\right)\right] \qquad (41)$$

7 Methodology

In the EIT reconstruction simulations our goal was to identify a conductive object within a circular non-conductive domain. We considered three cases where the object was placed in the center, between the center and the edge, and at the edge of the domain. These images, called ground-truth images, were created by using EIDORS with a mesh of 415 finite elements and 16 electrodes. Figure 2 shows the ground-truth images used in the experiments.

For all methods here discussed we used Eq. (36) as the objective function, where each dimension of the vector (solution candidate) corresponds to a particular finite element on the mesh. The stop criterion utilized was the maximum number of iterations.

The parameter used in the PSO for the experiments were: 'c1'and 'c2' = 2; ω (inertia weight) = 0.8; number of iterations = 500; and number of particles = 100.

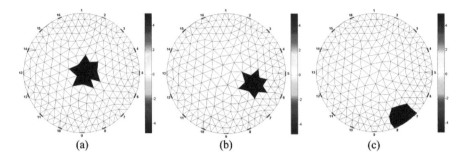

Fig. 2 Ground-truth images for the object placed (**a**) in the center (**b**), between the center and the edge, and (**c**) at the edge of the circular domain

For the DE' implementations, the agent referred to by the DE-SA algorithm is represented by a numerical vector containing the internal conductivity values of a candidate for the solution, that is the agent is a candidate for solving the TIE problem. The parameters used to implement the DE-SA method were initial number of agents: 100, probability of crossover: 90%, initial temperature: 200,000, and number of iterations: 500.

The parameters used for FSS' methods were 100 fish, $W_0 = 100$, $step_{ind_i} = 0.01$, $step_{ind_f} = 0.0001$, $step_{volt} = 2step_{ind}$ and $iterations = 500$. Whereas for dFSS' methods they were $\rho = 0.3$, $step_{init} = 0.01$, $decay_{min} = 0.999$, $decay_{max_{init}} = 0.99$, $decay_{max_{end}} = 0.95$, and $T_{max} = 500$.

8 Experimental Results and Discussion

In this section, the results obtained by reconstruction of EIT images through the use of hybrid methods previously described are presented.

8.1 Particle Swarm Optimization

The PSO and Gauss–Newton/PSO reconstructed images are shown in this section. They are presented in two results categories, which are:

- Qualitative: Figures 3 and 4 are the reconstructed images for the three ground-truths (center, between the center and the border, and the border) for the canonical PSO and the same technique with a particle generated by the Gauss–Newton algorithm inserted at the beginning of the iterative process, respectively.
- Quantitative: The graphs, shown in Figs. 5, 6, and 7 are the evolution of the relative error of the best particle, calculated by the objective function presented in Sect. 5.1, along the 500 iterations of each of the three executions of the algorithms

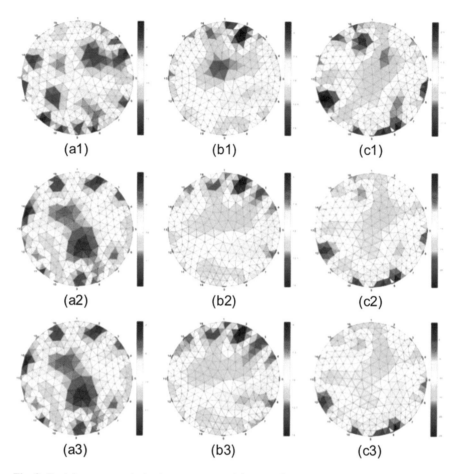

Fig. 3 Particle swarm optimization reconstructed images. Center ground-truth: 'a'; between the center and the border ground-truth: 'b'; and border ground-truth: 'c'. The numbers at the right side of the chars (1, 2, and 3) stand for 50, 300, and 500 iterations, respectively

for the center, between the center and the border, and the border ground-truth images, respectively.

8.1.1 Particle Swarm Optimization and with Non-blind Search Qualitative Discussion

As with the results, their discussion is also qualitatively and quantitatively separated.

The qualitative analysis considers images reconstructed and several aspects of the similarity with their respective ground-truth (Fig. 2) and the noise presence and cleanness of the circular domain.

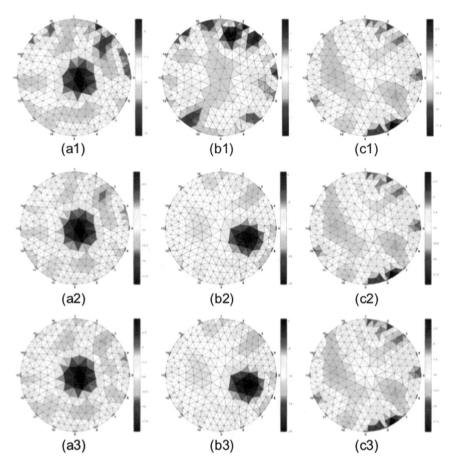

Fig. 4 Particle swarm optimization with non-blind search reconstructed images. Center ground-truth: 'a'; between the center and the border ground-truth: 'b'; and border ground-truth: 'c'. The numbers at the right side of the chars (1, 2, or 3) stand for 50, 300, and 500 iterations, respectively

The images generated by the PSO in Fig. 3 are generally noisier than the ones generated by the hybrid technique (PSO-NBS). It is important to clarify that by noisier we mean a less isolated resistive area (in red) with several artifacts around it. That means that the technique could not generate images with good isolation of the searched object (ground-truth). The same happened for the three ground-truth configurations.

On the other hand, the images generated by the PSO with non-blind search (Fig. 4) are cleaner than the former ones (by isolating the red object, which is the ground-truth). This factor means that the inclusion of a particle with prior knowledge, generated by the Gauss–Newton algorithm, can improve significantly the quality of the generated images.

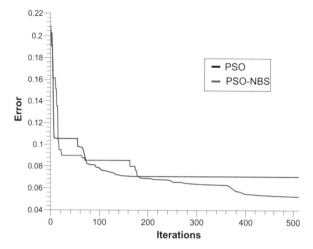

Fig. 5 Relative error for the ground-truth center placed

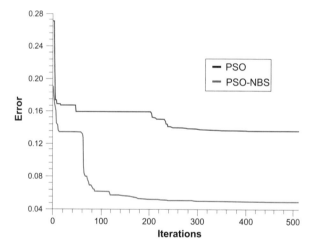

Fig. 6 Relative error for the ground-truth placed between the center and the border

In these experiments, the only drawback of the hybrid approach was when the image to be reconstructed was placed at the border of the domain (c1, c2, and c3). The reconstructed resistive area (in red) was slightly different from its ground-truth (Fig. 2 circular domain 'c'). Nevertheless, these images are still in better shape and with a better noise level than the ones generated with only the Particle Swarm Algorithm.

Under this analysis, it is clear that, qualitatively, the PSO-NBS hybrid approach (Fig. 4) overcame the only PSO approach (Fig. 3) in all the aspects.

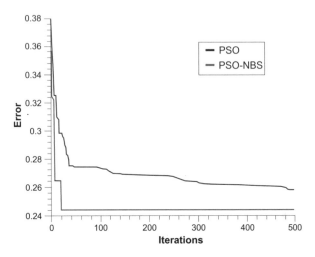

Fig. 7 Relative error for the ground-truth border placed

8.1.2 Particle Swarm Optimization and with Non-blind Search Quantitative Discussion

The quantitative analysis is based on the capacity of finding a low relative error value and the capacity of escaping local minimals along the iterations. Those factors can be observed in the relative error plots.

Under the three configurations (Figs. 5, 6 and 7), the hybrid approach (PSO-NBS, in red) shows a capacity of finding deeper values (i.e. lower relative error values). This probably happens because of the guidance of the search when a Gauss–Newton generated particle is put into the swarm. Besides, it is possible to notice that, unlike the PSO (in black), the PSO-NBS, in most of the cases, is able to escape local minimal regions. This can be seen by the linear trajectory of the back line in Fig. 5 around 150 iterations and in Fig. 6 around 340 iterations, when the error stops falling. The PSO-GN also has this trend of stagnating the error's fall; however, before that happens, this technique had already found a lower error than the PSO in the three cases.

The reconstruction problem regarding the border reconstruction images by the hybrid approach is also present in the quantitative results. In Fig. 7, the PSO-NBS (in red) stops its evolution (i.e. relative error falling) near the 20th iteration, which characterizes a local minimal stack, and therefore generation of a non-clear isolation of the searched object, as seen in Fig. 7, sub-units c1, c2, and c3. That is also the reason for those three images being equal, as there was no improvement (finding a better image) after the 20th iteration.

8.2 Differential Evolution Hybridized with Simulated Annealing

The results obtained for the hybridized reconstruction method are compared with the classical methods of DE and SA. The pseudocode and the characteristics of these methods were presented in Sect. 6.1.

It is important to remember that the initial agents were defined as having a random internal conductivity distribution in the range [0, 1] in the DE-SA method.

The results in Figs. 8 and 9 show images obtained from the DE and hybrid DE-SA method for an isolated object located in the center, between the center and the edge, and at the edge.

Where the situation represented in (a)–(c) is the location of an object in the center, between the center and the edge, and near the edge, respectively. (a1), (a2), and (a3) represent the images of the best candidates for solution of the object in the center in 50, 300, and 500 iterations, respectively. (b1), (b2), and (b3) represent the image of the best candidate for solution of the object between the center and the edge in 50, 300, and 500 iterations. Represents the image of the best candidate for solution to the object in near the edge in 50, 300, and 500 iterations.

Comparing the images given in Figs. 8 and 9, one can say that DE-SA can identify the objects with only 50 iterations; on the other hand, DE is able to identify only the case where the object is placed at the edge. Indeed, DE-SA obtained images are anatomically consistent and conclusive from 300 iterations for the three ground-truth images. Qualitatively, the DE-SA method showed high capacity in generating images with few artifacts from 300 iterations for all configurations, thus is a potential technique for eliminating image artifacts.

Figure 10 shows the graph for the error decrease in the function of the number of iterations for the DE and DESA methods. The curves in blue, red, and green represent the results for the object placed in the center, between the center and the edge, and at the edge, respectively, and the continuous and dotted lines represent the DE and DE-SA results, respectively.

From the graph in Fig. 10, we can observe that DE-SA always obtained lower results than DE, showing that the hybrid technique succeeds the non-hybrid technique. In this way, the comparison of the DE-SA method with the traditional method of DE shows that the former was more efficient in reconstruction, generating consistent images with a low relative noise level in the first 50 iterations. The reconstructions obtained with the DE-SA technique presented a greater edge definition when compared with the traditional method.

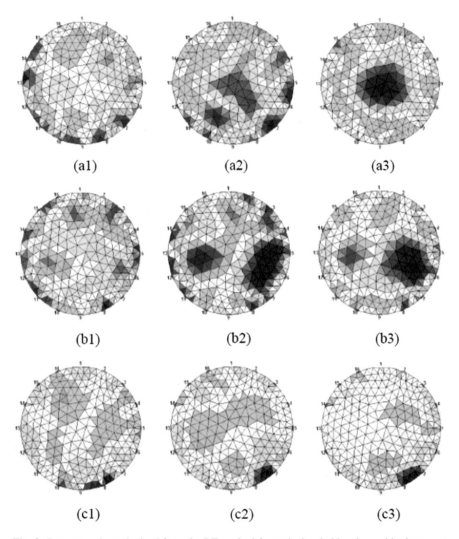

Fig. 8 Reconstructions obtained from the DE method for an isolated object located in the center (a1, a2, a3), between the center and the edge (b1, b2, b3), and near the edge (c1, c2, c3) of the circular domain for 50, 300, and 500 iterations, respectively

8.3   *Exponential Hybridization of the Fish School Search*

In this section the results for the hybridization of the FSS algorithm where the parameter $step_{ind}$ is updated following an exponential function as discussed in Sect. 6.2 will be presented.

In Figs. 11 and 12 reconstructed images are shown for the FSS and for the method hybridized with the decay of $step_{ind}$ exponentially, called FSSExp. When

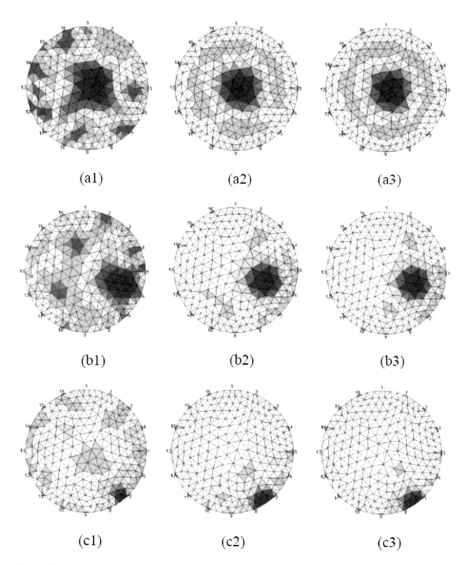

(a1)   (a2)   (a3)

(b1)   (b2)   (b3)

(c1)   (c2)   (c3)

Fig. 9 Reconstructions obtained from the hybrid DE-SA method for an isolated object located in the center (a1, a2, a3), between the center and the edge (b1, b2, b3), and near the edge (c1, c2, c3) of the circular domain for 50, 300, and 500 iterations, respectively

comparing the images, it is possible to observe that the FSS method was able to identify the object only when it is on the edge in 50 iterations, while the FSSExp appropriates the object at the edge and was also able to detect when the object is placed between the center and the edge. However, the results for 300 and 500 iterations show that the FSS method obtained images closer to the actual sized objects with less noise than FSSExp. The motivation to change the way the $step_{ind}$

Fig. 10 Graph of the value
of the objective function by
number of iterations for DE
and DE-SA

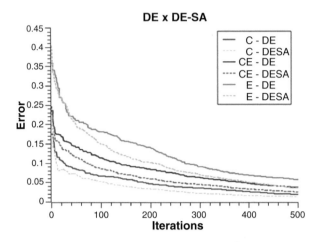

value decreases was the improvement between 300 and 500 iterations obtained by
the FSS method which is minimal and described by Fig. 11.

The data can be evaluated by the graph given in Fig. 13 quantitatively, where the
error is shown as a function of the number of iterations. The curves in shades of
blue, red, and green correspond to the images with the object in the center, between
the center and the border, and on the border, while the curves with the solid dotted
line show the results for FSS and FSSExp. The graph of Fig. 13 confirms what the
reconstructed images show. The first iterations show the curves very close; however,
the performance of the FSS ends up being higher than the FSSExp with the increase
in the number of iterations, perhaps the decay of the $step_{ind}$ parameter was very
aggressive, changing the exploratory search to exploitative too early. The graph
also shows the decrease in the value of the objective function obtained by the FSS
between 300 and 500 iterations.

8.4 Density Based on Fish School Search with Non-blind Search

In this section we will present the results obtained by the density based on FSS with
the non-blind search (dFSS + NBS) and compared with the method in its simplest
form (dFSS).

Figures 14 and 15 show reconstructed images by dFSS and dFSS+NBS, respectively. From these results, it is possible to observe that the use of the non-blind
search accelerated the search process for dFSS, as observed in the comparison of the
results in 50 iterations between the methods. In 50 iterations the dFSS obtained only
noisy and inconclusive images, while dFSS+NBS found the object in all three cases,
despite the high noise. In 500 iterations, the dFSS+NBS images have an object that

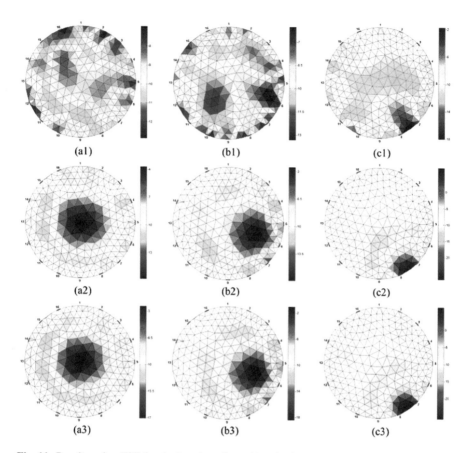

Fig. 11 Results using FSS for the location of an object in the center (a1, a2, a3), between the center and the edge (b1, b2, b3), and near the edge (c1, c2, c3) of the circular domain for 50, 300, and 500 iterations

is closer to the real one and have less noise when compared to the images obtained by the dFSS.

Figure 16 shows a graph that describes the fall of the relative error as a function of the number of iterations. The curves in shades of blue, red, and green correspond to the images with the object in the center, between the center and the border, and on the border, respectively, while the curves with the solid dotted line show the results for FSS and dFSS+NBS, respectively. From the graph, it is possible to observe that in the case of the object at the edge, the implementation of the non-blind search resulted quickly in the candidate with the best solution in the first iteration. For the other cases, the dFSS+NBS curve was higher than the dFSS. However, all curves of dFSS+NBS were lower than that of dFSS when the iterative process was over, demonstrating that the implementation of the solution from the Gauss–Newton method improved the performance of dFSS.

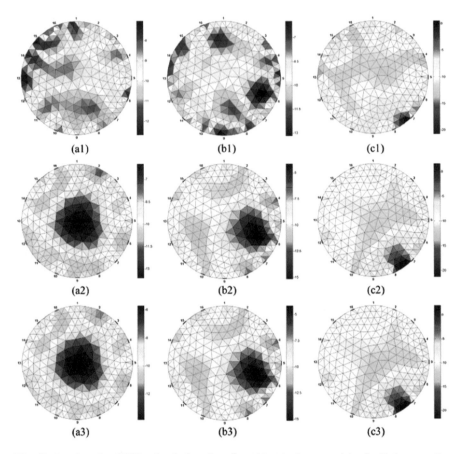

Fig. 12 Results using FSSExp for the location of an object in the center (a1, a2, a3), between the center and the edge (b1, b2, b3), and near the edge (c1, c2, c3) of the circular domain for 50, 300, and 500 iterations

9 Proposed Hardware Infrastructure

EIT images are acquired using the proposed prototype, whose main function is to control the injection of electric currents to pairs of electrodes and, afterwards, measure the resulting electric potentials; since all boundary potentials are acquired, a dedicated software is used to reconstruct images based on approximate numerical solutions [62]. All functions are organized by an embedded control system in the hardware that generates a data file; the system's architecture is shown in Fig. 17.

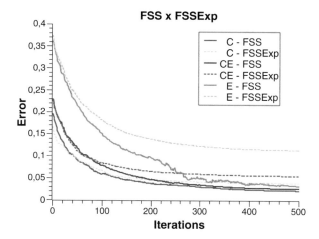

Fig. 13 Graph of the value of the objective function by number of iterations for FSS and FSSExp

- **Imaging area**: This is called a phantom and simulates a human tissue excited by only a little alternated current. The experimental environment has an electrolytic cell with an object immersed in liquid that is a normal saline solution (0.9% of NaCl), where 16 electrodes are distributed around the surface for the excitation and reading of electrical response potentials. The verification of the environmental impedances was made with a sensitive impedance meter during assembly for calibration of the device for capturing a signal in the order of millivolts.

- **Microcontrolled platform**: Based on low-cost open hardware, this is responsible for the general control system of the excitation module of the electrodes and also the reading of the voltages coming from the pairs to be considered, made by the multiplexing of analog inputs. The prototyping platform used was ARDUINO MEGA 2560, which offers many I/O pins, serial ports for programming and communication, and has a low purchase price ($10 on average). The control system of an EIT is developed in ARDUINO software IDE using a C language dialect.

- **Alternating current source**: A 1 mA_{pp} sine-wave source was dimensioned to meet the needs of a signal with low amplitude and frequencies in the range of 10–250 kHz. [63]. It is important to work with a low signal because currents injected into a human body can be dangerous, so research always must pay attention in this specific case.

- **16-Bit analog demultiplexer**: Responsible for the switching of the excitation signal through all the electrodes.

- **16-Bit analog multiplexer**: Provides to the microcontroller the voltage readings from a pair of electrodes following the techniques seen in the introduction.

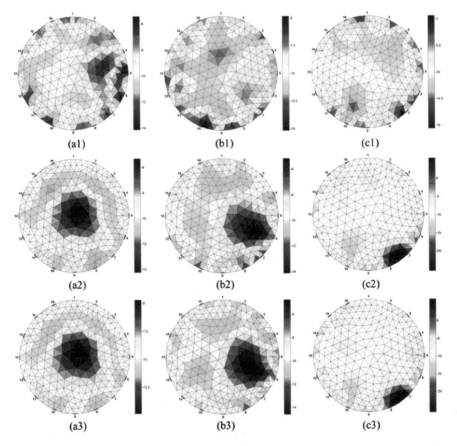

Fig. 14 dFSS results for an object placed in the center (a1, a2, a3), between the center and the edge (b1, b2, b3), and at the edge (c1, c2, c3) of the circular domain for 50, 300, and 500 iterations

- **Acquisition and pre-processing**: The signals collected on the electrodes through the multiplexers are treated and amplified for further reading before being converted to digital in the microcontrolled platform.
- **Computer communication**: Data from the reading are transmitted digitally in order to be processed by the reconstruction software, using a serial communication through USB port in Microcontrolled Platform.
- **Computational reconstruction**: In a computer, the impedance mapping data are processed by an algorithm that reconstructs the image with one of the proposed swarm intelligence optimizations.

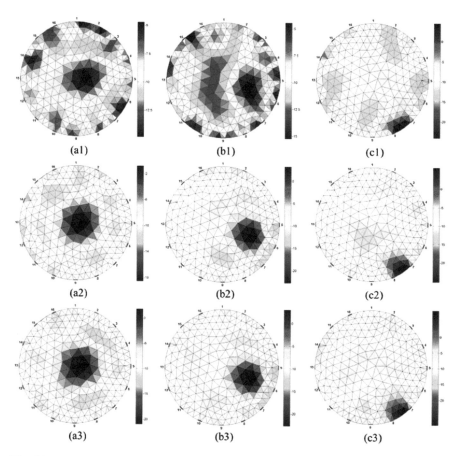

Fig. 15 dFSS+NBS results for an object placed in the center (a1, a2, a3), between the center and the edge (b1, b2, b3), and at the edge (c1, c2, c3) of the circular domain for 50, 300, and 500 iterations

9.1 The Embedded Control System

This activates the current signal on one electrode and stores the electrical potentials of the remaining electrode pairs, avoiding repetitions and readings of two equal electrodes. The software operation is described in Fig. 18 and shows the steps for reading the electrodes to prepare the data processing.

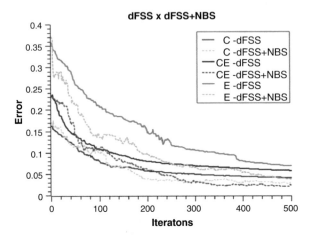

Fig. 16 Graph of the value of the objective function by number of iterations for dFSS and dFSS+NBS

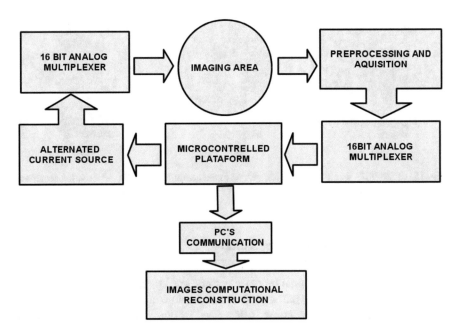

Fig. 17 Proposed hardware block diagram

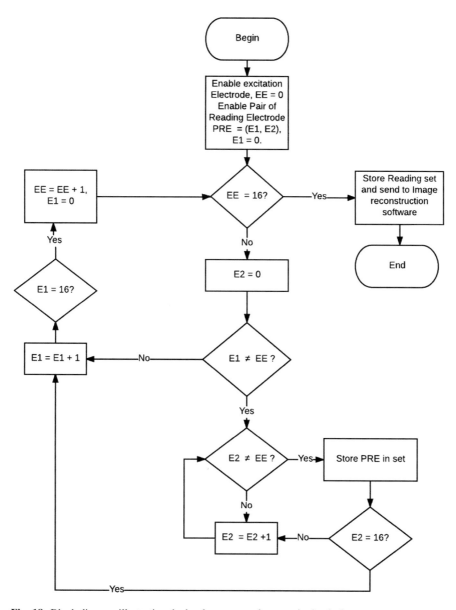

Fig. 18 Block diagram illustrating the hardware control system in the device

10 Conclusion

Regarding the PSO and its Gauss–Newton hybrid approach, it is clear that the hybridization has significantly improved the performance of the image reconstruction, both qualitatively and quantitatively. It is also important to highlight that neither the Gauss–Newton technique nor the PSO is sufficiently satisfactory when operating alone. This fact highlights the importance of the hybridization of these two techniques to perform a better technique for the EIT image reconstruction. The same can be observed in the implementation of the Gauss–Newton solution to the density based on the FSS algorithm. The hybridized technique outperformed the simple dFSS qualitatively and quantitatively.

One other hybridization described in this chapter was the implementation of SA in DE. The second technique has the capability to explore the search space; on the other hand the former is better able to exploit it. In this case, the hybridization aimed to improve the search process made by DE. The union of these techniques results in a better EIT reconstruction algorithm as was shown by the results obtained by the hybrid technique.

Finally the change of the linear decrease of the individual movement parameter to an exponential decrease in the FSS algorithm was considered. The goal was to increase the refined search during the iterative process; however, the strong decrease of the exponential made the change of exploration to exploitation search happen earlier. Thus, the hybridized method results were not better than the simple FSS results, as we expected.

References

1. C. Grosan, A. Abraham, Hybrid evolutionary algorithms: methodologies, architectures, and reviews, in *Hybrid Evolutionary Algorithms* (Springer, Berlin, 2007), pp. 1–17
2. Y. Wang, Z. Cai, G. Guo, Y. Zhou, Multiobjective optimization and hybrid evolutionary algorithm to solve constrained optimization problems. IEEE Trans. Syst. Man Cybern. B (Cybern.) **37**(3), 560–575 (2007)
3. C. Liang, Y. Huang, Y. Yang, A quay crane dynamic scheduling problem by hybrid evolutionary algorithm for berth allocation planning. Comput. Ind. Eng. **56**(3), 1021–1028 (2009)
4. F. Grimaccia, M. Mussetta, R.E. Zich, Genetical swarm optimization: self-adaptive hybrid evolutionary algorithm for electromagnetics. IEEE Trans. Antennas Propag. **55**(3), 781–785 (2007)
5. J.M. Peña, V. Robles, P. Larranaga, V. Herves, F. Rosales, M.S. Pérez, Ga-eda: hybrid evolutionary algorithm using genetic and estimation of distribution algorithms, in *International Conference on Industrial, Engineering and Other Applications of Applied Intelligent Systems* (Springer, Berlin, 2004), pp. 361–371
6. D.H. Wolpert, W.G. Macready et al., No free lunch theorems for search. Technical Report SFI-TR-95-02-010, Santa Fe Institute (1995)
7. D.H. Wolpert, W.G. Macready, No free lunch theorems for optimization. IEEE Trans. Evol. Comput. **1**(1), 67–82 (1997)
8. D.H. Wolpert, W.G. Macready, Coevolutionary free lunches. IEEE Trans. Evol. Comput. **9**(6), 721–735 (2005)

9. T. Cheng, B. Peng, Z. Lü, A hybrid evolutionary algorithm to solve the job shop scheduling problem. Ann. Oper. Res. **242**(2), 223–237 (2016)
10. P. Guo, W. Cheng, Y. Wang, Hybrid evolutionary algorithm with extreme machine learning fitness function evaluation for two-stage capacitated facility location problems. Expert Syst. Appl. **71**, 57–68 (2017)
11. R.R. Ribeiro, A.R. Feitosa, R.E. de Souza, W.P. dos Santos, A modified differential evolution algorithm for the reconstruction of electrical impedance tomography images, in *5th ISSNIP-IEEE Biosignals and Biorobotics Conference (2014): Biosignals and Robotics for Better and Safer Living (BRC)* (IEEE, New York, 2014), pp. 1–6
12. R.R. Ribeiro, A.R. Feitosa, R.E. de Souza, W.P. dos Santos, Reconstruction of electrical impedance tomography images using chaotic self-adaptive ring-topology differential evolution and genetic algorithms, in *2014 IEEE International Conference on Systems, Man, and Cybernetics (SMC)* (IEEE, New York, 2014), pp. 2605–2610
13. M.G. Rasteiro, R. Silva, F.A.P. Garcia, P. Faia, Electrical tomography: a review of configurations and applications to particulate processes. KONA Powder Part. J. **29**, 67–80 (2011)
14. G.L.C. Carosio, V. Rolnik, P. Seleghim Jr., Improving efficiency in electrical impedance tomography problem by hybrid parallel genetic algorithm and a priori information, in *Proceedings of the XXX Congresso Nacional de Matemática Aplicada e Computacional*, Florianopolis (2007)
15. F.C. Peters, L.P.S. Barra, A.C.C. Lemonge, Application of a hybrid optimization method for identification of steel reinforcement in concrete by electrical impedance tomography, in *2nd International Conference on Engineering Optimization* (2010)
16. V.P. Rolnik, P. Seleghim Jr., A specialized genetic algorithm for the electrical impedance tomography of two-phase flows. J. Braz. Soc. Mech. Sci. Eng. **28**(4), 378–389 (2006)
17. T.K. Bera, S.K. Biswas, K. Rajan, J. Nagaraju, Improving image quality in electrical impedance tomography (EIT) using projection error propagation-based regularization (PEPR) technique: a simulation study. J. Electr. Bioimpedance **2**(1), 2–12 (2011)
18. W.P. dos Santos, F.M. de Assis, Algoritmos dialéticos para inteligência computacional. Editora Universitária UFPE (2013)
19. A.E. Eiben, J.E. Smith, *Introduction to Evolutionary Computing*, vol. 2 (Springer, Berlin, 2015)
20. A.E. Eiben, M. Schoenauer, Evolutionary computing. Inf. Process. Lett. **82**(1), 1–6 (2002)
21. J. Kennedy, E. Russell, Particle swarm optimization, in *Proceedings of 1995 IEEE International Conference on Neural Networks* (1995), pp. 1942–1948
22. K.-L. Du, M.N.S. Swamy, *Bacterial Foraging Algorithm* (Springer International Publishing, Cham, 2016), pp. 217–225
23. C.J. Bastos Filho, F.B. de Lima Neto, A.J. Lins, A.I. Nascimento, M.P. Lima, A novel search algorithm based on fish school behavior, in *IEEE International Conference on Systems, Man and Cybernetics, SMC 2008* (IEEE, New York, 2008), pp. 2646–2651
24. S.S. Madeiro, F.B. de Lima-Neto, C.J.A. Bastos-Filho, E.M. do Nascimento Figueiredo, Density as the segregation mechanism in fish school search for multimodal optimization problems, in *Advances in Swarm Intelligence* (Springer, Berlin, 2011), pp. 563–572
25. A. Chikhalikar, A. Darade, Swarm intelligence techniques: comparative study of ACO and BCO. Self **4**, 5 (1995)
26. D. Karaboga, An idea based on honey bee swarm for numerical optimization. Technical report-TR06, Erciyes University, Engineering Faculty, Computer Engineering Department (2005)
27. S.-M. Chen, A. Sarosh, Y.-F. Dong, Simulated annealing based artificial bee colony algorithm for global numerical optimization. Appl. Math. Comput. **219**(8), 3575–3589 (2012)
28. S. Kirkpatrick, C.D. Gelatt, M.P. Vecchi et al., Optimization by simulated annealing. Science **220**(4598), 671–680 (1983)
29. T. de Castro Martins, M.d.S.G. Tsuzuki, Electrical impedance tomography reconstruction through simulated annealing with total least square error as objective function, in *2012 Annual International Conference of the IEEE Engineering in Medicine and Biology Society (EMBC)* (IEEE, New York, 2012), pp. 1518–1521
30. J. G. Sauer, Abordagem de Evolução diferencial híbrida com busca local aplicada ao problema do caixeiro viajante. PhD thesis, Pontifícia Universidade Católica do Paraná, 2007

31. R. Storn, K. Price, Differential evolution–a simple and efficient heuristic for global optimization over continuous spaces. J. Glob. Optim. **11**(4), 341–359 (1997)
32. G.T.d.S. Oliveira et al., Estudo e aplicações da evolução diferencial (2006)
33. S. Das, P.N. Suganthan, Differential evolution: a survey of the state-of-the-art. IEEE Trans. Evol. Comput. **15**(1), 4–31 (2011)
34. S. Das, A. Konar, Automatic image pixel clustering with an improved differential evolution. Appl. Soft Comput. **9**(1), 226–236 (2009)
35. C.J. Ter Braak, A Markov chain Monte Carlo version of the genetic algorithm differential evolution: easy bayesian computing for real parameter spaces. Stat. Comput. **16**(3), 239–249 (2006)
36. W.P. dos Santos, F.M. de Assis, Optimization based on dialectics, in *International Joint Conference on Neural Networks, IJCNN 2009* (IEEE, New York, 2009), pp. 2804–2811
37. K. Price, R.M. Storn, J.A. Lampinen, *Differential Evolution: A Practical Approach to Global Optimization* (Springer Science & Business Media, New York, 2006)
38. M.G.P. de Lacerda, F.B. de Lima Neto, A new heuristic of fish school segregation for multi-solution optimization of multimodal problems, in *Second International Conference on Intelligent Systems and Applications (INTELLI 2013)* (2013), pp. 115–121
39. A. Lins, C.J. Bastos-Filho, D.N. Nascimento, M.A.O. Junior, F.B. de Lima-Neto, Analysis of the performance of the fish school search algorithm running in graphic processing units, in *Theory and New Applications of Swarm Intelligence* (2012), pp. 17–32
40. A. Adler, T. Dai, W.R.B. Lionheart, Temporal image reconstruction in electrical impedance tomography. Physiol. Meas. **28**(7), S1 (2007)
41. A.R.S. Feitosa, Reconstrução de imagens de tomografia por impedância elétrica utilizando o método dialético de otimização. Master's thesis, Universidade Federal de Pernambuco (2015)
42. V.P. Rolnik, P. Seleghim Jr., A specialized genetic algorithm for the electrical impedance tomography of two-phase flows. J. Braz. Soc. Mech. Sci. Eng. **28**(4), 378–389 (2006)
43. M.G. Rasteiro, R.C.C. Silva, F.A.P. Garcia, P.M. Faia, Electrical tomography: a review of configurations and applications to particulate processes. KONA Powder Part. J. **29**, 67–80 (2011)
44. M. Cheney, D. Isaacson, J.C. Newell, Electrical impedance tomography. SIAM Rev. **41**(1), 85–101 (1999)
45. O.H. Menin, Método dos elementos de contorno para tomografia de impedância elétrica. Master's thesis, Faculdade de Filosofia, Ciências e Letras de Ribeirão Preto da Universidade de São Paulo (2009)
46. J.N. Tehrani, C. Jin, A. McEwan, A. van Schaik, A comparison between compressed sensing algorithms in electrical impedance tomography, in *2010 Annual International Conference of the IEEE Engineering in Medicine and Biology* (IEEE, New York, 2010), pp. 3109–3112
47. S.P. Kumar, N. Sriraam, P. Benakop, B. Jinaga, Reconstruction of brain electrical impedance tomography images using particle swarm optimization, in *2010 5th International Conference on Industrial and Information Systems* (IEEE, New York, 2010), pp. 339–342
48. T.J. Yorkey, J.G. Webster, W.J. Tompkins, Comparing reconstruction algorithms for electrical impedance tomography. IEEE Trans. Biomed. Eng. **34**(11), 843–852 (1987)
49. R. Bayford, Bioimpedance tomography (electrical impedance tomography). Annu. Rev. Biomed. Eng. **8**, 63–91 (2006)
50. H. Wang, G. Xu, S. Zhang, W. Yan, An implementation of generalized back projection algorithm for the 2-D anisotropic EIT problem. IEEE Trans. Magn. **51**(3), 1–4 (2015)
51. C.N. Lara Herrera, Algoritmo de tomografia por impedância elétrica baseado em simulated annealing. Master's thesis, Universidade de São Paulo (2007)
52. G.V. Momenté, B.H.L.N. Peixoto, M.S.G. Tsuzuki, T.D.C. Martins, New objective function for electrical image tomography reconstruction, in *ABCM Symposium Series in Mechatronics* (2013)
53. P. Hua, E.J. Woo, J.G. Webster, W.J. Tompkins, Finite element modeling of electrode-skin contact impedance in electrical impedance tomography. IEEE Trans. Biomed. Eng. **40**, 335–343 (1993)

54. V.P. Rolnik, Contribuição ao desenvolvimento de uma nova técnica de reconstrução tomográfica para sondas de visualização direta. PhD thesis, Escola de Engenharia de São Carlos da Universidade de São Paulo (2003)
55. M. Tang, W. Wang, J. Wheeler, M. McCormick, X. Dong, The number of electrodes and basis functions in EIT image reconstruction. Physiol. Meas. **23**(1), 129 (2002)
56. C.-N. Huang, F.-M. Yu, H.-Y. Chung, The scanning data collection strategy for enhancing the quality of electrical impedance tomography. IEEE Trans. Instrum. Meas. **57**(6), 1193–1198 (2008)
57. V.A.F. Barbosa, R.R. Ribeiro, A.R.S. Feitosa, V.L.B.A. da Silva, A.D.D. Rocha, R.C. Freitas, R.E. de Souza, W.P. dos Santos, Reconstrução de imagens de tomografia por impedância elétrica usando cardume de peixes, busca não-cega e algoritmo genético, in *Anais do 12 Congresso Brasileiro de Inteligência Computacional, ABRICOM, Curitiba, PR*, ed. by C.J.A. Bastos Filho, A.R. Pozo, H.S. Lopes (2015), pp. 1–6
58. A. Adler, W.R. Lionheart, Uses and abuses of EIDORS: an extensible software base for EIT. Physiol. Meas. **27**(5), S25 (2006)
59. Y. Liu, F. Sun, A fast differential evolution algorithm using k-nearest neighbour predictor. Expert Syst. Appl. **38**(4), 4254–4258 (2011)
60. K. Liu, X. Du, L. Kang, Differential evolution algorithm based on simulated annealing, in *Advances in Computation and Intelligence* (2007), pp. 120–126
61. P. Wang, X. Qian, Y. Zhou, N. Li, A novel differential evolution algorithm based on simulated annealing, in *2010 Chinese Control and Decision Conference (CCDC)* (IEEE, New York, 2010), pp. 7–10
62. M.F.M. Vallejo, C.N.L. Herrera, F.S. de Moura, J.C.C. Aya, R.G. Lima, The use of linear programming as search method of images in electrical impedance tomography, in *Proceedings of the 19th International Congress of Mechanical Engineering* (2007)
63. G. Singh, S. Anand, B. Lall, A. Srivastava, V. Singh, Development of a microcontroller based electrical impedance tomography system, in *2015 IEEE Long Island Systems, Applications and Technology Conference (LISAT)* (IEEE, New York, 2015), pp. 1–4

Index

accuracy, 10, 14, 20, 21, 23–28
active contour models, 145, 147, 148
adaptive threshold, 199
ant colony optimization, 118
artificial bee colony, 91, 118
artificial neural network, 3, 15, 89
attractiveness probability, 129

bacterial foraging algorithm, 118
bat algorithm, 117
Baum-Welch, 91, 92, 99, 100
Baum's re-estimation algorithm, 92
β-hill climbing, 91
binarization, 194
binary chromosome, 38
blob detector, 96

Cambridge hand dataset, 94
cellular automata, 33
chromosome, 1, 4, 7–9, 11–13, 17, 19, 24, 27
chromosome design, 60
cluster, 15–18, 22, 24, 27, 28
code book, 89
color-based segmentation, 88
complex system, 33
computer vision technology, 88
conditional random field, 88
confusion matrix, 10, 13, 14, 21, 23
continuous domain, 163, 164, 168
convergence criterion, 193, 194

cooling factor, 129
cooperative method, 19, 21
crossover, 8, 9, 11–14, 16, 19, 27, 28
cuckoo search algorithm, 91, 92, 94, 103, 104, 117
cyber glove, 89

decision tree, 93
deep convolutional networks, 117
deformable model, 146, 147
defuzzification, 126
density based on fish school search, 221
    food sharing, 221
    operator
        collective-instinctive movement, 222
        collective-volitional movement, 223
        memory, 222
        partitioning, 222
desaturation, 95
design of the proposed GA-SVM model, 62
differential evaluation algorithm, 119, 128
differential evolution, 91, 190, 215
    crossover, 217
    initialization, 216
    mutation, 216
    selection, 217
    strategies, 218
distance-based classification, 58
DRIVE database, 163, 176, 179, 180
dynamic, 4, 11, 12, 14, 18, 28

© Springer International Publishing AG, part of Springer Nature 2018
S. Bhattacharyya (ed.), *Hybrid Metaheuristics for Image Analysis*,
https://doi.org/10.1007/978-3-319-77625-5

Printed in the United States
By Bookmasters